Francis Sansom, William Darton

Flora londinensis

Francis Sansom, William Darton

Flora londinensis

ISBN/EAN: 9783742891716

Manufactured in Europe, USA, Canada, Australia, Japa

Cover: Foto ©berggeist007 / pixelio.de

Manufactured and distributed by brebook publishing software
(www.brebook.com)

Francis Sansom, William Darton

Flora londinensis

INDEX I.

In which the Plants contained in the fifth Fasciculus are arranged according
to the System of LINNÆUS.

LIGUSTRUM VULGARE. PRIVET or PRIM.

LIGUSTRUM *Lin. Gen. Pl.* DIANDRIA MONOGYNIA.

Cor. 4 fida. *Bacca tetrasperma.*

Raii Syn. RAMOSAE BACCIFERAE.

LIGUSTRUM ——— *Lin. Syst. Vegetab.* p. 52. *Sp. Pl.* p. 10. *Fl. Suec. n. 3. Haller. Hist.* n. 530. *Scopol. Flor. Carniol. n. 4. Hudson. Fl. Angl.* ed. 2. p. 3. *Lightfoot Fl. Scot.* p. 72.

LIGUSTRUM Germanicum. *Bauh. Pin.* 475. *Ger. em.* p. 1394. *Parkinson.* p. 1446. *Raii Syn.* p. 465. Privet or Prim.

FRUTEX ————

FOLIA opposita, ————

FLORES albi, odorati, paniculati.
PANICULA terminalis, densa, subpyramidata.

RAMI ————

CALYX: ————

COROLLA ————

STAMINA: FILAMENTA duo, opposita, brevissima, alba. ANTHERAE ———— erectae, longitudine fere corollae. POLLEN flavescens, *fig.* 3.

PISTILLUM: GERMEN ———— STYLUS filiformis, ———— STIGMA ————

PERICARPIUM: BACCA globosa, glabra, nigra, unilocularis, *fig.* 5.
SEMINA ————

A SHRUB, ufually about fix feet high, branched, the bark of a greenish-ash colour, irregularly fprinkled with numerous prominent points: branches oppofite, the young ones flexible and purplish.

LEAVES oppofite, ftanding on very fhort foot-ftalks, ovate-lanceolate, fmooth on each fide, perfectly entire, the lower ones at the bottoms of the fmall branches ——.

FLOWERS white, fweet-fcented, forming a panicle.
PANICLE about two inches in length, clofe and fomewhat pyramidal.

BRANCHES of the panicle, as well as the flowerftalks, ruddier when magnified.

CALYX a ———— of one leaf, very fmall, hemifpherical, and whitifh, the vefel having four teeth, which are upright and very minute, *fig.* 1.

COROLLA of one petal, funnel-fhaped, white, foon changing to a reddifh-brown colour. The *tube* cylindrical, longer than the calyx. *Limb* deeply divided into four fegments, which are ————, ovate, thick, and obtufe, *fig.* 2.

STAMINA: two FILAMENTS, oppofite, very fhort and white. ANTHERAE rather large, upright, almoft the length of the corolla. POLLEN yellowifh, *fig.* 3.

PISTILLUM: GERMEN roundifh. STYLE filiform, white, a little thickened above. STIGMA obtufe, thickifh, fcarce perceptibly bifid, *fig.* 4.

SEED-VESSEL a round, fmooth, fhining, black, berry of one cavity, *fig.* 5.
SEEDS three or four, convex on one fide, and angular on the other, *fig.* 6.

Previous to the publication of the *Flora Japonica* by Profeffor THUNBERG *, the prefent celebrated fucceffor to the immortal LINNAEUS, Botanifts were acquainted with one fpecies of Liguftrum only. That gentleman defcribes another, to which he gives the name of *japonicum*, and characterifes the two in the following manner:

Liguftrum vulgare foliis ovatis obtufis, panicula fingulatim trichotoma.
In japonica: panicula foliis ovatis acuminatis punctulis decuffatis trichotoma.

It is of utility, not to its ornament, for of our Englifh or even foreign fhrubs exceed the common Privet. It is ufed alfo to form fuch hedges as are required in the dividing of gardens for fhelter or ornament; the luftre or ever-green Privet, as it is called, which is wide a variety of the common fpecies, is ufually preferred for this purpofe. The Privet bears clipping admirably well: is not liable to be deftroyed by infects, and having roots formed only of fibres, it robs the ground lefs than almoft any other fhrub. It is found to thrive better in the infide of great cities than moft others; in that whoever has a little garden in fuch places, and is defirous of having a few plants that ———— grow and healthy, may be gratified in the Privet, because it will flourifh and look well there. Mice as lays it will grow well under the fhade and drip of trees.

The laft mode of raifing Privet is from feeds, though it is capable of being propagated by layers and cuttings.

The Privet is not apt to be eaten by cattle, and the *Sphinx Ligustri*, or *Privet Hawk Moth*, one of the largeft as well as the moft beautiful infects we have, is almoft the only one that feeds on it in its Caterpillar ftate. There are few gardens having Privet in which this Caterpillar may not be found in the months of Auguft and September. The readieft way of difcovering it is by its dung, which is fufficiently vifible under thofe fhrubs on which it feeds. The *Meloe vesicatoria*, commonly known by the name of Cantharides, or Blifter-beetle, is found alfo on the leaves of this fhrub. The leaves of the Privet continue on the plant till fpring advances, and in times of fcarcity are eaten by different forts of birds, but by none with fo much avidity as the *Bulfinch* (*Loxia Pyrrhula*). Birdcatchers who know this, often catch them in the following manner: they take fome large boughs of the Privet in berry, ftick them into the ground where Bulfinches frequent, lime the top twigs, and place a call bird underneath. The berries are alfo used in ———— colouring of wines, and as affording a purple colour to ftain prints; but for thofe feveral purpofes they are much better materials in common ufe.

It ufually grows in woods and hedges, is not nice in its foil or fituation, but flourifhes moft in a moift foil; flowers in July, and ripens its berries in Autumn.

It is found with three forts at a plant, with variegated leaves, and white berries. HALLER.

* Caroli Petri Thunberg Flora Japonica, Lipfiae 1784.

Ligustrum vulgare.

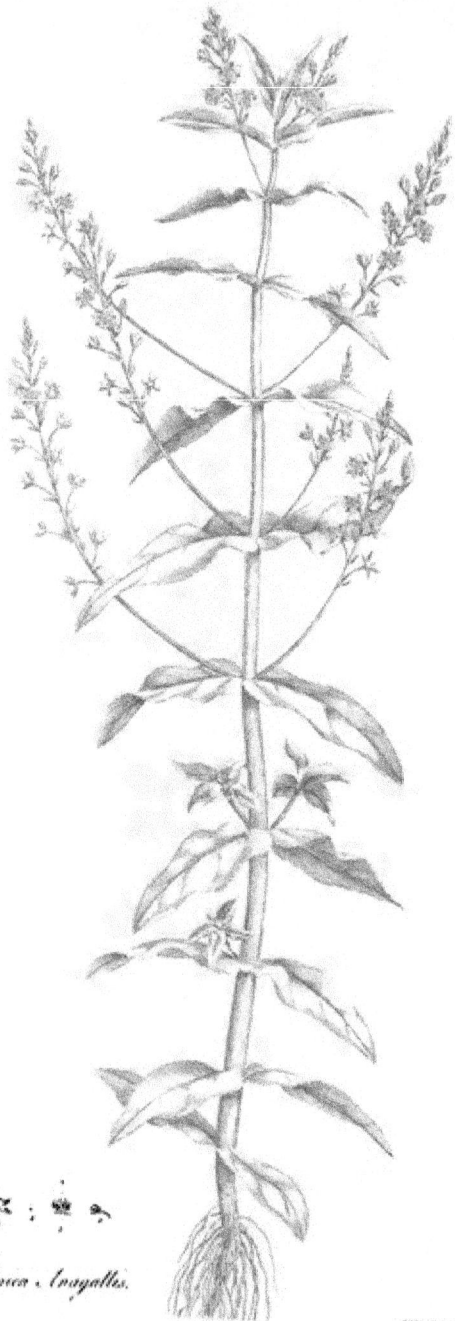

Veronica Anagallis.

VERONICA ANAGALLIS. WATER SPEEDWELL.

VERONICA *Lin. Gen. Pl.* Diandria Monogynia.

Cor. Limbo 4-partito, lacinia infima angustiore. Capsula bilocularis.

Raii Syn. Gen. 18. Herbæ fructu sicco singulari flore monopetalo.

VERONICA Anagallis racemis lateralibus, foliis lanceolatis serratis, caule erecto. *Lin. Syst. Vegetab. p.* 56. *Sp. Pl. p.* 16. *Fl. Suec. n.* 13.

VERONICA foliis lanceolatis serratis, glabris, ex alis racemosa. *Haller hist. n.* 553.

VERONICA Anagallis Scopoli *Fl. Carn. n.* 12.

ANAGALLIS aquatica maior folio oblongo. *Bauh. Pin.* 252.

ANAGALLIS aquatica folio oblongo crenato. *Park.* 1237.

ANAGALLIS aquatica maior. *Ger. emac.* 620.

VERONICA aquatica longifolia media. *Raii Syn.* 280. The Middle Long-leav'd Water Speedwell or Brooklime. *Hudson, Fl. Angl. ed.* 2. *p.* 5. *Lightfoot Fl. Scot. p.* 73.

RADIX annua, fibrosa.	ROOT annual, and fibrous.
CAULIS erectus, pedalis ad bipedalem, teres, subangulosus, glaber, ad basin usque ramosus, inferne purpurascens.	STALK upright, from one to two feet high, round, slightly angular, smooth, branched quite to the bottom, below purplish.
FOLIA opposita, sessilia, lanceolata, sæpe ovato-lanceolata, serrata, glabra, venosa, pallide viridia.	LEAVES opposite, sessile, lanceolate, often ovato-lanceolate, serrated, smooth, veiny, of a pale green colour.
FLORES racemosi, numerosi, triginta quadraginta aut etiam plures in singulo racemo.	FLOWERS growing in racemi, numerous, from thirty to forty, or even more on one racemus.
RACEMI laterales, oppositi, longissimi, suberecti.	RACEMI lateral, opposite, very long, nearly upright.
PEDUNCULI ad lentem subvifcidi.	FLOWER-STALKS somewhat viscid when magnified.
BRACTEÆ lanceolatæ.	FLORAL-LEAVES lanceolate.
CALYX: Perianthium quadripartitum, persistens, laciniæ ovato-lanceolatæ, acutæ, lævibus, trinervibus, subæqualibus, *fig.* 1.	CALYX: a Perianthium deeply divided into four segments, and permanent, the segments ovato-lanceolate, pointed, smooth, three ribbed, and nearly equal, *fig.* 1.
COROLLA monopetala, rotata, pallide purpurea, lacinia superiore et duabus lateralibus venis faturatioribus striatis, *fig.* 2.	COROLLA monopetalous, and wheel-shaped, of a pale purple colour, the uppermost segment and the two lateral ones streaked with deeper veins of the same colour, *fig.* 2.
STAMINA: Filamenta duo, purpurascentia, medio crassiora; Antheræ concolores; Pollen album, *fig.* 3.	STAMINA: two Filaments of a purplish colour, thickest in the middle; Antheræ of the same colour; Pollen white, *fig.* 3.
PISTILLUM: Germen viride; Stylus declinatus, purpurascens, superne crassior; Stigma obtusum, *fig.* 4.	PISTILLUM: Germen green; Stylus depending, purplish, thickened above; Stigma blunt, *fig.* 4.
PERICARPIUM: Capsula bilocularis, subinde trilocularis, subrotunda, vix emarginata, polysperma, *fig.* 5.	SEED-VESSEL: a Capsule of two cavities, sometimes three, roundish, scarcely emarginate, containing many seeds, *fig.* 5.
SEMINA platima, subrotunda, minutissima, *fig.* 6.	SEEDS numerous, roundish, and very minute, *fig.* 6.

The *Veronica Anagallis* is a much more general plant than the *Scutellata*, being found in almost every watery ditch, but especially in those which communicate with the Thames, on the edges of which it is also extremely common.

It is apt to vary considerably according to situation; when it grows in ditches that have a considerable depth of water, it becomes much taller, the stalk is proportionably thicker, and the leaves are apt to be curled; when it grows out of the water, the plant is smaller, the leaves are broader, flatter, and of a paler hue; in all situations its racemi are remarkably long and full of flowers, and its seeds are uncommonly small and numerous.

It blossoms from June to September.

The seed-vessels are sometimes found very much enlarged; on cutting them open a small larva was found in some, and a pupa in others, which, on being kept a proper time, produced a small Oestrus or Worm.

VERONICA SCUTELLATA. BOG SPEEDWELL.

VERONICA *Lin. Gen. Pl.* DIANDRIA MONOGYNIA.

Cor. Limbo 4-partito, lacinia infima angustiore. *Capsula* bilocularis.

Raii Syn. Gen. 18. HERBÆ FRUCTU SICCO SINGULARI FLORE MONOPETALO.

VERONICA *scutellata* racemis lateralibus alternis: pedicellis pendulis, foliis linearibus integerrimis. *Lin. Syst. Vegetab. p.* 57. *Sp. Pl. p.* 16. *Fl. Suec. n.* 17.

VERONICA foliis lanceolatis, serratis, glabris, ex alis racemosa. *Haller Hist.* 533.

VERONICA *scutellata. Scopoli Fl. Carn. n.* 22.

ANAGALLIS aquatica angustifolia scutellata. *Bauh. Pin.* 252.

VERONICA aquatica angustifolia minor. Narrow-leav'd Water Speedwell, or Brooklime. *Raii Syn. p.* 280. *Hudson. Fl. Angl. ed.* 2. *p.* 5. *Lightfoot Fl. Scot. p.* 74.

RADIX perennis, fibrosa, fusca.

ROOT perennial, fibrous, of a brown colour.

CAULIS: paulo supra terram fasciculi plerumque flexilis erumpunt, qui humi repunt, caulis floridus suberectus, debilis, teres, vix angulosus, glaber, ramosus, sempedalis ad podalem, basi etiam aliquando repens.

STALK: just above the ground young shoots spring forth, which are for the most part debilitate of flowers and creep on the earth, the flowering stalk is nearly upright, weak, round, scarce perceptibly angular, branched, from six inches to a foot in height, sometimes also creeping at bottom.

FOLIA opposita, sessilia, lineari-lanceolata, glabra, minutius et rarius dentata.

LEAVES opposite, sessile, betwixt linear and lanceolate, smooth, finely tooth'd, seth distant.

FLORES albi, seu pallide carnei, racemosi.

FLOWERS white, or of a pale flesh colour, growing in racemi.

RACEMI laterales, plerumque alterni, laxi, flexuosi, multiflori.

RACEMI lateral, for the most part alternate, loose, crooked, and bearing many flowers.

BRACTEÆ minutæ, lanceolatæ.

FLORAL-LEAVES minute, and lanceolate.

PEDUNCULI capillares, alterni, demum penduli.

FLOWER-STALKS capillary, alternate, finally pendulous,

CALYX: PERIANTHIUM parvum, quadripartitum, laciniis ovato-lanceolatis, subæqualibus, *fig.* 1.

CALYX: a PERIANTHIUM small, deeply divided into four segments, which are ovato-lanceolate and nearly equal, *fig.* 1.

COROLLA monopetala, rotata, plerumque alba, laciniis superiore venis purpureis picta, *fig.* 2.

COROLLA monopetalous, wheel-shaped, for the most part white, the upper segment streaked with purple veins, *fig.* 2.

STAMINA: FILAMENTA duo, medio incrassata, alba; ANTHERÆ albæ, *fig.* 3.

STAMINA: two FILAMENTS, thickest in the middle, white; ANTHERÆ white, *fig.* 3.

PISTILLUM: GERMEN viride; STYLUS declinatus, albus; STIGMA obtusum, flavescens, *fig.* 4.

PISTILLUM: GERMEN green; STYLE depending, white; STIGMA blunt, yellowish, *fig.* 4.

PERICARPIUM: CAPSULA compressa, suborbiculata, emarginata, bilocularis, polysperma, ad 16. *fig.* 5.

SEED-VESSEL a CAPSULE nearly round, flattened, emarginate, of two cavities, containing numerous seeds, to 16. *fig.* 5.

SEMINA orbiculata, plana, flava, *fig.* 6.

SEEDS round, flat, and yellow, *fig.* 6.

This species of Veronica is distinguished from the others by several characters, such as, its place of growth, which is peculiar, it being seldom found but on bogs, or the edges of ponds, especially such as we find on heaths and moors, hence we have called it *Bog Speedwell*; the narrowness as well as smoothness of its leaves also strikingly distinguishes it; LINNÆUS's term of *integerrimis*, as applied to them, is certainly too strong, for they are always toothed, though faintly, and in a singular manner; and if these characters were not sufficient, the loose straggling manner in which the flower stalks grow, would at once point out the *Scutellata* as a distinct species.

It is common in the situations above described on most of our heaths, and flowers from June to September.

Veronica scutellata.

VALERIANA LOCUSTA. CORN SALLAD.

VALERIANA *Lin. Gen. Pl.* TRIANDRIA MONOGYNIA.

          Cal. o. *Cor.* 1-petala, basi hinc gibba, supera. *Sem.* 1.

VALERIANA *Locusta* floribus triandris, caule dichotomo, foliis linearibus. *Lin. Syst. Vegetab. p.* 73. *Sp. Pl. p.* 47. *Fl. Suec. n.* 36.

VALERIANA foliis oblongis, rarius incisis, corona seminis simplici, acuminata. *Haller Hist.* 214.

VALERIANA *Locusta. Scopoli Fl. Carn. n.* 46.

VALERIANA campestris inodora major. *Bauh. Pin.* 165.

VALERIANELLA arvensis praecox humilis semine compresso. *Mor. Hist.*

LACTUCA agnina. *Ger. emac.* 310. *Park.* 812. *Raii Syn. p.* 201. Lamb's-Lettuce or Corn-Sallet. *Hudso. Fl. Angl. ed.* 2. *p.* 13. *Lightfoot Fl. Scot. p.* 85.

RADIX annua, fibrosa, pallide fusca.

CAULIS erectus, spithamaeus, pedalis et ultra, pro ratione loci, teres, angulato-striatus, subpubescens, tener, ad utrum latus lineas purpurascens, dichotomus.

FOLIA radicalia, plurima, patentissima, subsucculenta, glabra, venosa, subrugosa, obsolete dentata, caulina opposita, sessilia, remota, ad basin praecipue ciliata, subcrebra, suprema subserrata.

FLORES minimi, caerulescentes, corymbosi.

CALYX nullus.

COROLLA longitudine germinis, tubulosa, faciniolaces, quinquefida, laciniis rotundatis, patentibus, subaequalibus, *fig.* 1.

STAMINA: FILAMENTA tria, alba, longitudine corollae. ANTHERAE parvae, albae, *fig.* 2.

PISTILLUM: GERMEN inferum, nodum, majusculum, obovatum, viride, utrinque fisse exarcum, hinc convexum, subgibberium, inde planiusculum, *fig.* 4. STYLUS staminibus paulo brevior. STIGMA trifidum, *fig.* 3.

SEMINA plurima, nuda, pallide fusca, subrotunda, acutiuscula, parum compressa, transversim rugosa, *fig.* 5.

ROOT annual, fibrous, of a pale brown colour.

STALK upright, from four inches to a foot or more in height, according to its place of growth, round, grooved or angular, slightly downy, tender, usually purplish on one side, dichotomous.

LEAVES next the root numerous, somewhat spreading, slightly succulent, smooth, veiny, a little wrinkled, inversely ovate, faintly toothed, those of the stalk opposite, sessile, remote, at the base particularly, edged with hairs, somewhat upright, the uppermost ones slightly serrated.

FLOWERS very minute, of a bluish colour, growing in a corymbus.

CALYX wanting.

COROLLA the length of the germen, tubular, faintly violet-coloured, divided into five segments, which are roundish, spreading, and nearly equal, *fig.* 1.

STAMINA: these FILAMENTS of a white colour, the length of the corolla. ANTHERAE small and white, *fig.* 2.

PISTILLUM: GERMEN placed below the corolla, naked, rather large, inversely ovate, green, having a narrow groove on each side, convex and somewhat gibbous on one side, flattish on the other, *fig.* 4. STYLE a little shorter than the stamina. STIGMA trifid, *fig.* 3.

SEEDS numerous, naked, of a pale brown colour, roundish, a little pointed, somewhat flattened, and transversely wrinkled, *fig.* 5.

In treating of the *Valeriana dioica* we had occasion to notice the extreme inconstancy of the fructification in this genus; an inconstancy scarcely to be paralleled in any other tribe, and affecting not only the Linnaean system, as depending on number of stamina, but such systems also as may be founded on the form of the corolla, or structure of the seed. In the officinalis, dioica, and several other valerians, the seeds are furnished with a pappus or down, here they are altogether naked.

The present plant is a well known culinary one; the radical leaves are in general use in the spring to mix with other sallad herbs, and sometimes eaten alone: the French call them *Salad de Preter*, from their being generally eaten in Lent.

It grows wild in corn fields, on walls, banks, and in gardens. In corn-fields it is usually very small, grows with a single stem, and often occurs with diseased heads, occasioned by some insect. The leaves are sometimes more than usually serrated. A variety of this sort is made a species of by RAY. There are several other varieties mentioned by LINNAEUS in his *Species Plantarum*, which have not come under our observation.

It flowers in May, and ripens its seed in June.

Valeriana Locusta.

Alopecurus
pratensis.

Alopecurus pratensis. Meadow Foxtail-Grass.

ALOPECURUS *Lin. Gen. Pl.* Triandria Digynia.

Cal. 2-valvis *Cor.* 1-valvis.

Raii Syn. Gen. 27. Herbæ graminifoliæ flore imperfecto culmiferæ.

ALOPECURUS *pratensis* culmo spicato erecto, glumis villosis, corollis muticis. *Lin. Syst. Vegetab.* p. 93. *Sp. Pl.* p. 88. *Fl. Suec.* 20.

ALOPECURUS spica ovata. *Haller. Hist. n.* 1539.

GRAMEN phalaroides majus sive italicum. *Bauh. pin.* 4.

GRAMEN alopecuroides majus. *Ger. emac.* 10.

GRAMEN phalaroides majus. *Parkins.* 1164.

GRAMEN alopecuro simile glabrum cum pilis longiusculis in spica unocordata mihi denominatum. *I. B. II. Raii Syn.* p. 396. The most common Foxtail-grass. *Hudson. Fl. Angl. ed. 2.* p. 27. *Lightfoot Fl. Scot.* p. 91. *Schreb. Gram.* 133. t. 19. f. 1.

RADIX perennis, fibrosa, fibris pallidè fuscis.

CULMI sesquipedales, bipedales, et haud infrequenter tripedales, erecti, teretes, striati, læves, ad basin purpurei, radicantes.

FOLIA palmaria, seu spithamea, sensim in acutum mucronem terminata, glabra, striata, parte superna et ad margines si digiti deorsum ducantur aspera, sutum unam cum dimidia communiter aut duas sere latæ. Vaginæ striatæ, læves, in superiore parte culmi inflatæ. Membranæ brevis, obtusa.

SPICA sesquiuncialis, biuncialis, duas etiam nonnunquam cum dimidia uncias longa, duas tresque lineas lata, teres, cylindracea, obtusa, mollis.

SPICULÆ unifloræ, compressæ, utrinque ciliatæ, reniformes, unocronato-tridentatæ, *fig.* 1.

CALYX: Gluma bivalvis, uniflora, valvula subæqualibus, ovato-lanceolatis, concavis, compressis, triperviis, nervis pilosis, *fig.* 2.

COROLLA univalvis, valvula concava, longitudine calycis, alruia, subdiaphana, superne nervis tribus viridibus insignita, subbarbata; arista calyce duplo sere longiore, basio valvulæ versus basin inserta, *fig.* 3.

STAMINA: Filamenta tria, capillaria. Antheræ oblongæ, utrinque bifurcæ, plerumque purpurascentes, demum serrugineæ, *fig.* 4.

PISTILLUM: Germen ovatum, minimum. Styli duo, villosi, reflexi, calyce longiores. Stigmata simplicia, *fig.* 5.

SEMEN ovatum, minimum, glumis tectum, *fig.* 6, 7.

ROOT perennial and fibrous, the fibres of a pale brown colour.

STALKS a foot and a half, two feet, and not unfrequently three feet high, upright, round, finely grooved, smooth, at bottom purple, and tillering.

LEAVES a hand's breadth or sheer span in length, gradually tapering to a point, smooth, striated, if drawn backward across the fingers feeling rough on the upper side and on the edges, commonly a line and a half or almost two in breadth. Sheaths striated, smooth, on the upper part of the stalk inflated. Membrane short and blunt.

SPIKE an inch and a half, two inches, and sometimes even two inches and a half long, and two or three lines broad, round, cylindrical, blunt and soft.

SPICULÆ one flower in each, flat, each side edged with hairs, ribbed, slightly tridentate, the middle point longest, *fig.* 1.

CALYX: a Gluma of two valves, containing one flower, the valves nearly equal, ovate and pointed, flattened, three-ribbed, the ribs hairy, *fig.* 2.

COROLLA of one valve, the valve hollow, the length of the calyx, whitish, somewhat transparent, marked on the upper part with three green ribs, and bearded; the beard or awn almost as long again as the calyx, inserted into the back of the valve towards the base, *fig.* 3.

STAMINA: three capillary Filaments. Antheræ oblong, forked at each end, for the most part purplish, finally ferruginous, *fig.* 4.

PISTILLUM: Germen ovate, very minute. Styles two, villous, reflexed, longer than the glumes of the calyx. Stigmata simple, *fig.* 5.

SEED ovate, very minute, covered by the glumes, *fig.* 6, 7.

In a former number of this work, containing the *Fifteen Fescues*, we gave a copious extract from that excellent work on Grasses, the *Historie-long der Gräser* of Professor Schreber: we now present our readers with an abridged account from the same author of another grass, apparently of much greater consequence in agriculture.

The Meadow Foxtail-grass is chiefly an inhabitant of the northern part of our moderate zone, being found abundantly in most parts of Germany, Holland, France, England, Denmark, Norway, Sweden, and Russia. Professor Gmelin has also found it plentifully in Siberia.

Though the grasses in general are not so strongly attached to particular locations as many plants are, yet they are always more abundant, and superior in goodness, in some one kind of ground than another. The Meadow Foxtail loves a meadow ground somewhat low, and moderately wet, with a good soil, though it will also grow in dry, and even in quite wet ground; yet, in the wet, it remains free, small, and disappears by little and little, while, in the latter, other grasses are apt to overpower and supplant it.

In such districts of Norway as are celebrated for the goodness of their meadows, it always makes a considerable part of the hay; and the same remark has been made by Mr. Stillingfleet and Professor Kalm in England, respecting the best meadows about London.

The Meadow Foxtail is one of those grasses which appear first in the spring, and sometimes blow twice in the same year. In respect to flowering, it sufficeth to notice that it commences at the abovementioned information. In Cheshire it puts forth its silvery spikes about the beginning of May; when the seed is ripe, which with us takes place before hay-making; the spike remains unchanged in its shape for some time; the little husks containing the seed may easily be stripped off, but fall off very slowly of themselves.

Experience proves that the Meadow Foxtail grass has a power of vegetating quickly. Its shoots proceed with such vigour, that it may very well be cut three times in a year. Its stalks are strong, and provided with large leaves, which are soft and juicy. Their taste is as that of good fodder-grass ought to be, sweetish and agreeable, having, when made into hay, neither the hardness of straw, nor the roughness or unpleasant taste attendant on some of the other grasses; we may therefore consider it as heading the first place among the good grasses, either used as fresh fodder, or made into hay, especially for the larger cattle. Though the sheep in such meadows as abound with this grass, do not improve in the flavour of their wool, yet they give a preference to it, both green and dried. On the whole, we may with truth assert, that hay is better in proportion to the quantity of Meadow Foxtail-grass there is among it; not to mention that such hay has the advantage in the weight, and consequently goes farther than hay made of the finer grasses.

In the northern countries, Sweden especially, the meadows are frequently laid waste by a small destructive caterpillar, which produces a moth called, by Linnæus, Phalæna graminis: it has been discovered, that the Alopecurus pratensis remains untouched by this destructive insect; so far, therefore, from injuring this grass, it gives it an opportunity, by weakening and destroying the others, to extend itself farther; but though its particular calls on forward growth exempts it from the ravages of this species of caterpillar, there is another which is particularly fond of it, viz. the Phalæna potatoria, yet as this feeds largely on its foliage, and never increases greatly, it suffers little from it.

As this grass, therefore, appears to be our author of so much consequence in the making and improving of meadows and pastures, he proceeds to give some account how this improvement may be effected.

In this business the first thing of moment, he observes, is the necessary choice and preparation of the ground; if that be in the power of the cultivator, and as the Meadow Foxtail is found neither to thrive in a soil that is quite dry, or quite wet, he prefers a wet one rendered moderately dry by draining.

After procuring a piece of ground naturally fit, or rendered so by art, he recommends it to be ploughed up immediately after harvest, before the wet season sets in, in which state it is to remain all the winter; the frost breaking the clods, renders it fit for sowing in the spring, at which time you must throw in your seeds of the Meadow Foxtail, mixed with other proper pasture herbs, together with a crop of oats; the latter, when sufficiently grown, may be cut for fodder.

A meadow, thus improved, requires all the care necessary in the management of meadows; in particular, a copious watering after hay-making, if the season prove unusually dry, must not be omitted. If after some years the soil should become bound, or noxious plants increase in such a manner as to make the meadow less productive, which often happens when the soil or situation is unfavourable, the meadow must be broken up and fresh sown.

The procuring of the seed, requisite even for a tolerably large sowing, is attended with but little difficulty, if we can only get some slips or roots of this grass. The great number of seeds which grow upon one spike, of which more than one spring from each slip; the double crop in one summer, and the rapid growth of this grass, evince this sufficiently. The gathering of the seed itself is very easy; it needs only to be stripped off with the hand, and put in a bag, and if there be a large quantity together, spread out and dried, even the hay-seed of such meadows as abound with Meadow Foxtail is useful in sowing; but we must well observe how it is mixed: good hay-seed should contain a greater proportion of grass-seeds than of other herbs; the latter must be esculent and nutritive, without any mixture of hard, woody, or succulent ones, which corrupt the hay; much less should it contain tasteless, acrid, or poisonous plants. But it may be asked, where is such hay-seed to be obtained? Certainly the meadows are rare which contain a mixture of proper plants unadulterated with noxious ones; hence the best method will be to collect separately the seeds of the most useful grasses and meadow plants, to increase them singly, to compound the hay-seed of them, and to sow therewith, at least, small meadows, from whence we may, in process of time, obtain a sufficient stock of seed for a more general cultivation.

---

* This disposition of grass to flower twice than once in the same year, is perhaps deserving of more attention than may have hitherto been paid to it. We have noticed it in two years strongly in the present grass, the yellow Oat, the red Oat, and some others; on the contrary, there have grass, viz. the Poa trivialis, slowly ripened, which we have never observed to flow the least tendency to throw up a flowering stem twice in the same year. While this may serve as an additional character, whereby it may be distinguished from the Poa trivialis, it may also recommend it as a valuable grass for meadow purposes, where hints are established, and afford the eye. We observed, in treating of the Poa trivialis, that its own sort of the creeping kind, it will probably be found, that all these grasses which have this property.

† Its third state of flowering with us.

‡ In the neighbourhood of London, hay-making generally commences three or four weeks sooner than it does fifty miles from town. Whether this practice hath arisen from the richness of soil encouraging the growth of the herbage, or from the meadows abounding more with early grasses, it may perhaps be difficult to determine; but certainly, by this practice, as soap off the advantages from these early grasses which are led by longer time, and hence the seeds of our hay falls must be proportionably better than those at a distance, is very quick it producible as hay.

§ In the papers of the Bath Agricultural Society, vol. X. p. 59, the Rev. Mr. Swayne, of Pucklechurch, in Gloucestershire, gives an account of a very curious insect, which, feeding within the husks of the spikes, renders them barren, or kills the seeds. "On cutting open the husks, when I judged the seed to be approaching to ripeness, I found almost every husk either occupied by a small substance, of a deep yellow or orange colour, or with caterpillars feeding. On applying the microscope, this substance proved to be a congeries of microscopic, which being threshed out on a sheet of white paper, and separated from each other, displayed the exact shape and texture of little insects which are sometimes found to come out swarm, and which are known among husbandmen by the name of hoppers. The flies likewise, which these caterpillars produce, were found to be very like the hopper flies, only evidently smaller."

‖ We should prefer the beans and of things, or beginning of hayseason, for the purpose of sowing grass seeds, provided the season proved favourable.

¶ Should the barley intended to be laid down be very foul, we recommend, separate ploughings and harrowings, and this be more than once further, would be necessary. Farmers are divided in their opinions respecting the propriety of sowing Oats or barley with grass-seeds; from experience, that the oats close the pores; grass must have by rubbing it of its nourishment, that the thicker or better offer led therefore does it good.

Hebenstreitia geniculata.

ALOPECURUS GENICULATUS. JOINTED FOX-TAIL GRASS.

ALOPECURUS   *Lin. Gen. Pl.* TRIANDRIA DIGYNIA.

Cal. 2-valvis. *Cor.* 1-valvis.

Raii Syn. Gen. 27. HERBÆ GRAMINIFOLIÆ FLORE IMPERFECTO CULMIFERÆ.

ALOPECURUS *geniculatus culmo spicato infracto, corollis muticis, Lin. Syst. Vegetab. p. 93. Sp. Pl. 89. Fl. Suec. n. 60. Haller. hist. n. 1511.*

ALOPECURUS *geniculatus culmo adscendente, spica cylindrica, glumis apice divergentibus pilosis, Hudson Fl. Angl. ed. 2. p. 27.*

ALOPECURUS *geniculatus Scopol. Fl. Carn. n. 82.*

GRAMEN   aquaticum geniculatum spicatum. *Bauh. pin.* 3. *Scheuchz. Agrost.* 72.

GRAMEN   fluviatile spicatum. *Ger. emac.* 14.

GRAMEN   aquaticum spicatum. *Parkins.* 1277. *Raii Syn.* 396. Spiked Floce Grass. *Lightfoot, Fl. Scot. p. 90. Order N. Dan.* 564.

RADIX perennis, fibrosa, fibris albicantibus, et quandoque subfuscis.

CULMI plures, pedales, sesquipedales et ultra, inferne procumbentes, et sæpe reptantes, suberecti, geniculati, infracti, ramosi, superne nodi, striati, præsertim in solo arido plus minus bulbosi.

FOLIA duo aut tres lineas lata, striata, superne digitis eorsum ductis aspera, inferne lævia, superiora brevia, uncialia aut biuncialia, patentia, sæpe ad margines crispa, membrana ad basin folii ovata, acuta, ragræ læves, striatæ, ventricosæ.

SPICÆ unciales, sesquiunciales et ultra, subcylindraceæ, forma et colore maxime variantes, nunc obtusæ nunc ad apicem sensim attenuatæ, virescentes, purpurascentes, aut etiam nigricantes procul saltem visæ.

FLOSCULI imbricati.

CALYX: GLUMA uniflora, bivalvis, compressa, valvulis oblique truncatis, pubescentibus, trinerviis, carina ciliata, *fig.* 1.

COROLLA: GLUMA univalvis, oblonga, ovata, truncata, quinquenervis, pellucida, nuda, aristata, *fig.* 2. Arista juxta basin exserta corolla duplo longiore, *fig.* 3.

STAMINA: FILAMENTA tria, corolla longiora; ANTHERÆ oblongæ, primum purpureæ, demum ferrugineæ, *fig.* 4.

PISTILLUM: GERMEN subrotundum; STYLI duo, cirrhosi, albidi, extra calycem protensi, *fig.* 5.

ROOT perennial, fibrous, the fibres whitish, sometimes inclined to brown.

STALKS several, a foot, a foot and a half or more in length, below procumbent, and often creeping, nearly upright, jointed, crooked, above naked and striated, branched, the base especially in a dry soil more or less bulbous.

LEAVES two or three lines broad, striated, the upper side if drawn backwards betwixt the fingers rough, the under side smooth, the uppermost leaves short, an inch or two inches long, spreading, often crisped at the edges, the membrane at the base of the leaf, ovate and pointed, the sheaths smooth, striated, and bellying out.

SPIKE an inch, an inch and a half or more in length, somewhat cylindrical, varying greatly both in form and colour, sometimes blunt, and sometimes tapering to a point, greenish, purplish, and even blackish, at least when viewed at a distance.

FLORETS imbricated.

CALYX: a GLUME of two valves, containing one flower, flattened, the valves obliquely truncated, downy, three-ribbed, the keel ciliated, *fig.* 1.

COROLLA: a GLUME of one valve, oblong, ovate, truncated, five-ribbed, pellucid, without hairs, and bearded, *fig.* 2. the Beard or awn proceeding from near the base, and twice the length of the corolla, *fig.* 3.

STAMINA: three FILAMENTS, longer than the corolla; ANTHERÆ oblong, at first purple, afterwards ferruginous, *fig.* 4.

PISTILLUM: GERMEN roundish; STYLES two, slender, feathery, and extended beyond the calyx, *fig.* 5.

It is in the depressed parts of meadows, where water is occasionally apt to stagnate, that this species of Fox-Tail Grass particularly delights to grow, nor is it unfrequent on the edges of ponds, streams, and wet ditches, where it often makes its way into the water; it is also, though more rarely, found in dry pastures; and, according to these several situations, it is found to vary.

In the first, the stalks are procumbent at the base, spread themselves on the ground, and extend a foot or more in length; before they rise upwards, the spikes often assume a blackish or deep purple colour, which causes it to be noticed by the Farmer, who distinguishes it by the name of Black Grass*. In the second, it is very much enlarged in its size, and approaches near to the *Alopecurus pratensis*; but the stalk still retains towards the bottom its crooked appearance. In the third, it grows more upright, the spike becomes much slenderer, and the base of the stalk often swells out into a kind of bulb, as in the *Avena elatior*, and this variety has been called *Alopecurus bulbosus*; in all these several varieties, the geniculatus cannot easily be mistaken for any other species of *Alopecurus*.

It flowers in June.

Cattle eat it readily, nevertheless it cannot be recommended as a profitable Grass; nor do the more observing Farmers consider it as such: indeed, where such Grass is apt to abound, the best practice would be to fill up the depressions, and sow the ground with better Grasses.

---

* The Farmer also distinguishes the *Alopecurus agrestis* (*myosuroides, Fl. Lond.*) by the name of Black Grass.

Bromus giganteus.

BROMUS GIGANTEUS. TALL BROME GRASS.

BROMUS *Lin. Gen. Pl.* TRIANDRIA DIGYNIA.

Cal. a-valvis. Spicula oblonga, teres, disticha : arista infra apicem.

Raii Syn. Gen. 27. HERBÆ GRAMINIFOLIÆ FLORE IMPERFECTO CULMIFERÆ.

BROMUS *giganteus panicula nutante, spiculis quadrifloris :* aristis brevioribus, *Lin. Syst. Vegetab. p.* 103. *Spec. Plant. p.* 114. *Fl. Suec.* n. 34.

BROMUS *giganteus panicula ramosa nutante, ramis binatis, spiculis subquadrifloris arista brevioribus. Hudson Fl. Angl. p.* 51.

BROMUS *glaber, locustis quadrifloris nutantibus, aristis longissimis. Haller. hist.* n. 1510.

BROMUS *giganteus. Scopoli Fl. Carn.* n. 116. VAR. *glabra et minor.*

GRAMEN *bromoides aquaticum latifolium, panicula sparsa tenuissime aristata. Scheuchz. Agrost. p.* 264. *t.* 5. *fig.* 17.

GRAMEN *sylvaticum glabrum, panicula recurva. Vaill. Paris. p.* 93.

GRAMEN *avenaceum glabrum, panicula e spicis raris singulis composita, aristis tenuissimis. Raii hist. 1900. Syn. p.* 415. *Loeselsoed Fl. Sect. p.* 104.

RADIX perennis, fibrosa.	ROOT perennial and fibrous.
CULMUS tripedalis et ultra, erectus, lævis, geniculis plerumque purpureis.	STALK three feet or more in height, upright, smooth, the joints for the most part purple.
FOLIA semunciam lata, læte viridia, lævia, inferne nitida, basi appendiculis ex fusco purpureis utrinque, caulem amplexantibus instructa, pagina inferne scabriuscula, minime pilosa, superne glabra, membrana brevissima.	LEAVES half an inch broad, of a bright-green colour, smooth, shining underneath, furnished at the base on each side with two purplish-brown appendages, which embrace the stalk, beneath below a little rough to the touch, but not hairy, above smooth, the membrane very short.
PANICULA ampla, pedalis etiam, sparsa, ramis plerumque binatis, nutantibus, sensuelis, scabriusculis.	PANICLE large, even a foot long, loose, branches generally growing in pairs, all one way, drooping, and roughish.
SPICULÆ ovato-lanceolatæ, subquinquifloræ, semunciales, plerumque virides, lævæ, aristatæ: Aristæ albæ, spiculis paulo longiores, flexuolæ, scabræ.	SPICULÆ ovato-lanceolate, containing about five flowers, half an inch in length, for the most part green, smooth, and bearded : Beards white, a little longer than the spiculæ, crooked, and rough.
CALYX: GLUMA bivalvis, valvulis inæqualibus, acuminatis, viridibus, marginibus albidis, majore lanceolatior, minore unica subdiaphana rotaro, *fig.* 1.	CALYX: a GLUMA of two valves, the valves unequal, pointed, green, with white edges, the large valve marked with three, and the small one with one somewhat transparent line. *fig.* 1.
COROLLÆ GLUMA bivalvis, valvulis subæqualibus, viridibus, tenibus, margine albis, exteriore majore, concava, obsolete trinervis, aristata, ariftâ glumâ longiore paulo infra apicem exserta, interiore minore, planiuscula, albida, *fig.* 2, 3.	COROLLA a GLUME of two valves, the valves nearly equal, green, smooth, the edges white, the outer one largest, hollow, faintly three-ribbed, and bearded, the beard longer than the glume, and proceeding from a little below the point, the interior one least, somewhat flat and whitish, *fig.* 2, 3.
NECTARIUM: GLUMULÆ duæ, acuminatæ, ad basin germinis, *fig.* 4.	NECTARY: two small pointed GLUMES at the base of the germen, *fig.* 4.
STAMINA: FILAMENTA tria, capillaria, alba; ANTHERÆ flavæ, bifurcæ, *fig.* 5.	STAMINA: three capillary, white FILAMENTS; ANTHERÆ yellow and forked, *fig.* 5.
PISTILLUM: GERMEN obovatum, viride, stigmma, STYLI duo, patentes, ad basin usque ramosi, *fig.* 6. auct. *fig.* 7.	PISTILLUM: GERMEN inversely ovate, green and stigma, STYLES two, spreading and branched quite to the bottom, *fig.* 6, magnified, *fig.* 7.
SEMEN oblongum, ex nigro purpurascens, intra glumas adhærentes, inclusum, *fig.* 8, 9.	SEED oblong, of a blackish-purple colour, enclosed within the glumes which adhere to it, *fig.* 8, 9.

There is only one grass for which this species of *Bromus* is liable to be mistaken, and that is the *Bromus Asper*, already figured, they are both grasses, and grow in similar situations, indeed frequently together; they have been confounded by SCOPOLI, who makes the *Asper* a variety of the *giganteus*; but the least attention would have taught him, that they were materially different.

The sheath of the lower leaves in the *Asper* is covered with long stiff hairs, which are wanting in the *giganteus*: the leaves of the *giganteus* are glossy on the under side, and those of the *Asper*, near their extremities, appear as if a black ligature had been tied round them; but there is a character almost peculiar to this grass, the base of the leaf is terminated by two small appendages, of a reddish-brown colour, which closely embrace the stalk, and will never fail to distinguish it from the *Asper*; the spiculæ also, if no other distinguishing characters were present, would be all-sufficient, being shorter by almost one half, containing fewer flowers, and having aristæ or awns longer in proportion to the spiculæ and more crooked: we may add another character which we have discovered from cultivation, the *giganteus* is a perennial, whereas the *Asper* is only an annual or biennial, a circumstance which we were not sufficiently apprised of when we described that plant.

This grass is frequent enough in the neighbourhood of London, in woods, and under hedges, especially such as are accompanied by a wet ditch, nor is it uncommon by the sides of the Thames; the situation which it affects with us, is more agreeable to the name given it by SCHEUCHZER, than to the account delivered by LINNÆUS in his *Species plantarum*, where he says, Habitat in Europæ *silvis siccis*: we very rarely or never find it in meadows; hence, though a productive grass, there seems not much probability of its becoming a good grass for meadows or pastures.

It flowers from July to September.

Holcus mollis.

HOLCUS MOLLIS. CREEPING SOFT-GRASS.

HOLCUS *Lin. Gen. Pl.* POLYGAMIA MONOECIA.

HERMAPHROD. *Cal.* Gluma 1-fl. 2-flora. *Cor.* Gluma aristata. *Stam.* 3.
Styli 2. *Sem.* 1.

MASC. *Cal.* Gluma 2-valvis. *Cor.* o. *Stam.* 3.

HOLCUS *mollis* calyce aristato, rhachi villosa, flosculo hermaphrodito mutico: masculo aristâ geniculatâ.
Lin. Syst. Veget. p. 910. Sp. Pl. p. 1485.

GRAMEN *cannorum longius radicatum majus et minus. Bauh. Pin. 1.*

GRAMEN paniculatum molle, calice gramineo cauleo aristato. *Morif. Hift. 3. p. 202.*

GRAMEN miliaceum paniculatum molle. *Raii Hift.* 1385. *Scheuchz. Agroft. p. 155. Vaill. Parif. p. 87.*

GRAMEN miliaceum aristatum molle. *Ray Syn. p. 404. Hudson. Fl. Angl. ed. 2. p.* 440. *Lightfoot Fl. Scot. p. 631. Schreb. Agroft. l. 20.*

RADIX perennis, rushi cauda bullæ repens.	ROOT perennial, creeping like the garden couch-grass.
CULMI longiusculi et vitrei, tenues erecti, foliosi, nodosi, geniculis albis, laeviis, culmi etiam flexiles decumbunt ad terram rough reclinati, foliis crebrioribus, alternis, lanceolatis, vestiti.	STALKS a foot and a half or more in length, most commonly upright, leafy, jointed, the joints white and woolly, stems also trile prolonging no spikes, inclined more to the ground, and covered with more numerous, alternate, lanceolate, leaves.
FOLIA ad tres vel quatuor lineas lata, molli villo pubescentes, circum nervi et bafin toti alba, siccata, vaginæ striatæ, foliorumata, sulcata.	LEAVES three or four lines in breadth, covered with soft short hairs, the membrane at the base of the leaf white and clasp, the sheath striated, furrowed at based and sulcated.
PANICULA blanch, erecta, inftante antheſi diffusa, demum conferta.	PANICLE two inches in length, upright, during the flowering spread out, afterwards closed up.
RAMULI pediculis purpurascentes, pilosi.	BRANCHES of the panicle purplish and hairy.
SPICULÆ bisloræ etiam triflorae, fig. 3. a. albotae flofculos purpurascentes, tinctulo nonnihil hermaphrodito.	SPICULÆ containing two, sometimes three flowers, fig. 3. a. whitish, or slightly tinged with purple, all the flowers hermaphrodite.
CALYX: gluma bivalvis, utrinque ciliata, ceterarum aridu, valvula altera minior et paulo longiore, trinerve, nervis obscure viridibus, fig. 1. 2.	CALYX: a gluma of two valves, edged on both sides with hairs, colourable naked, one of the valves larger and a little longer than the other, having three ribs, of an obscure green colour, fig. 1. 2.
COROLLA bivalvis, valvulis longitudine fubaequalibus, sed pilosis, viridibus, extériore majore, glabra, gibbula, interiore plana et latiore fubnervali, bisplddis, e dorfo majoris valvulæ superioris floreuli extergit arista spicula longire prime recta, demum tortilis, geniculata, fig. 3. 4.	COROLLA of two valves, the valves nearly equal in length, hairy at bottom, of a green colour, the outermost largest, smooth, and gibbous, the innermost flat, broadest ribbed where magnified, and a little striged, from the back of the largest valve of the uppermost flower arises an awn, longer than the spicula, at first straight, lastly twisted and bent, fig. 3. 4.
STAMINA. FILAMENTA tria, capillaria, ANTHERÆ oblongae, flavae, utrinque biferae, fig. 5.	STAMINA: three capillary FILAMENTS. ANTHERÆ oblong, yellow, forked at each end, fig. 5.
PISTILLUM: GERMEN fubrotundum, oblium, minimum. STYLI duo, plumosi, fig. 6.	PISTILLUM: GERMEN roundish, oblong, very small. STYLES two, feathery, fig. 6.
NECTARIUM: glandula duæ, lanceolatæ, ad bafin germinis, fig. 7.	NECTARY: two glands, lanceolate glands at the base of the germen, fig. 7.
SEMINA duo, nuda, ovato-acuta, altera ariſtata, altera mutica, glumis calycinis inclusis, fig. 8.	SEEDS two, naked, ovate, pointed, the one bearded, the other naked, inclosed within the glumes of the calyx, fig. 8.

Notwithstanding this grass has been well named and described by some of the older Botanists, particularly MORISON and RAY, its characters do not appear to be generally well understood. BauH. Hallerus considers it as too nearly related to the *lanatus*, to be with propriety considered as a distinct species; and Mr. LIGHTFOOT, in his *Flora Scotica*, entertains similar doubts.

We have cultivated the two in separate beds, close to each other, for several years; have noticed them with a marked attention, where they have grown wild; and, from a variety of characters, are led to consider them as perfectly distinct.

The most striking of these characters we shall here enumerate. In the first place they differ vitaly in their natural places of growth; while the *lanatus* is most commonly found in meadows and pastures, the *mollis* rarely occurs but in woods and its environs. We have, indeed, frequently found the *lanatus*, which is by far the most general grass of the two, in a wood; but we never recollect faing the *mollis* in meadows or pastures, and ftill rarely in corn-fields, where it has been faid chiefly to grow. Corn West is in particular affords a strong inftance of its attachment to shady fituations. Contrary to what some authors affert, we have even found the *mollis* the leaft plant; or, if it has been observed equally till as the other, it has produced by far the moft fcanty panicle; nor do the spicula, in general, affume that brilliant colour which to eminently diftinguishes those of the *lanatus* on their firft coming out. But the character which puts its being a fpecies out of all doubt, is its root: that of the *lanatus* does not creep, while the *mollis* poffeffes that property in a degree equal to the ſtrongeſt couch grass. The other characters which ſtrikingly diftinguiſh that species are its woolly joints and its large pointed fpicula, in which the beard, or awn, is invariably much larger than the glumes of the calyx.

In speaking of the *lanatus* we took notice of the impropriety of separating that grass from the general grass, because one of the flowers in each fpecula was imperfect*. The fructifications of the prefent fpecies are yet more ſtrongly for its union with the others: here both flowers are hermaphrodite, both have ſtamina and featheres ſtyles, and both produce apparently perfect seeds. Indeed we can perceive no characters to diftinguiſh it from an *Aira*, to which genus it perhaps with propriety belongs.

SCHREBER's figure gives a good repreſentation of the panicle when cloſed, but neither repreſents the joints or root well.

As we conſider the *Holcus lanatus*, which is much to be preferred to the preſent ſpecies, as a very indifferent grass for cattle, ſo we cannot but look on the *mollis* as one of the worſt ſpecies of couch; and, if it ſhould ever become a practice to ſow certain woods with grass feeds, this ſpecies ought ſurely to be eradicated.

It flowers in July.

* SCHREBER, from a circumſtance of this fort, has in our opinion clearly enough placed the *Aira* *cristata* with the *Holcus*.

Hordeum murinum

HORDEUM MURINUM. WALL BARLEY.

HORDEUM *Lin. Gen. Pl.* Triandria Digynia.

Cal. lateralis, bivalvis, uniflorus, ternus.

Reň Sp. Gen. 27. Herba gramine folia, flore compacto cylindrica.

HORDEUM *murinum* flosculis lateralibus masculis aristatis, involucris intermediis ciliatis. *Lin. Syst. Vegetab. p.* 108. *Sp. Pl. p.* 126. *Fl. Suec. n.* 113.

HORDEUM spicis crassis, longis aristatis, calycinis glumis aristatis. *Haller Hist. n.* 1338.

HORDEUM *murinum.* *Scopoli Fl. Carn. n.* 1241.

GRAMEN hordaceum minus et vulgare. *Bauh. Pin. 6.*

HORDEUM spurium vulgare. *Parkinson* 1147.

GRAMEN hordeaceum et locula spicula. *Ger. emac. 79. 2 ed Ste. p. 59.* Wild Rie or Rie-Grass. Wall-Barley, Way-Bennet. *Bauh. Fl. Angl. 22 & p. 58. Linnaeus Fl. Suec. p. 108.*

RADIX annua, fibrosa, albida vel fulvidea.

CULMI plures, pedales et sesquipedales, subereti, foliosi, basi procumbentes, patulo, geniculati, graminis majusculi, gelidioribus.

FOLIA palmaria in explicatione etiam sex uncias longa, duas vel tres lineas lata, longitate, mollipube vestita, basi appendiculis duabus albis, acuminatis, amplexicaulibus, indurata, membrana brevissima, obtusa; vagina vix pubescens.

SPICÆ palmares, et ultra, parum nutantes, pallide virentes, compressae, spicis basibs diductis basi abducis.

CALYX: Involucrum hexaphyllum, triflorum, foliolis setaceis, acuminatis, aristis corollæ breviorbus, duobis, duobus intermediis basi latiorbus, ciliatis, *fig.* 1.

FLOS intermedius hermaphroditus, laterales masculi, omnibus crispatis et forma Graillare, *fig.* 2. *Flos Hermaphrod.*

COROLLA bivalvis, valvula exterior oblongo ovata, acuminata, edidens muricula, bases, schipunctum aristum bimaculatum terminat, *fig.* 4. valvula interior lanceolata, plana, medio sulcato, apice emarginato truncata, *fig.* 3. ad basin exterioris hujus valvulæ exterior aristat corolla longitudinis filamentorum, *fig.* 8.

NECTARIUM: Glumulae duæ, acuminatæ ad basin geminam, *fig.* 7.
STAMINA: Filamenta tria, capillaria, plurali corollæ radice breviora. Antherae parvæ, e flavo virescentes, *fig.* 5.
PISTILLUM: Germen ovatum, pubescens. Stylla duo, villosi, *fig.* 6.

ROOT annual, fibrous, whitish or of a brownish colour.
STALKS numerous, a foot or a foot and a half high, nearly upright, leafy, procumbent at the base, and crooked or bended, jointed, the joints rather larger and paler than the stalk.

LEAVES a hand's breadth or in some even six inches in length, and two or three lines broad, somewhat glaucous, and covered with a soft down, furnished at the base with two small, white, pointed appendages, which embrace the stalk; membrane very short and obtuse; sheath scarcely downy.

SPIKES a hand's-breadth or more in length, drooping a little, of a pale green colour, flat, and not unlike those of common barley.

CALYX: an Involucrum of six leaves, containing three flowers, the leaves running out to a long bristly point, shorter than the beards of the corolla, the two intermediate ones broader at the base than the others, and edged with hairs, *fig.* 1.

FLOWER in the middle hermaphrodite, the side ones males, all alike bristly and sharp, *fig.* 2. *Hermaphrodite Flower.*

COROLLA of two valves, the outer valve oblong-ovate, with a long point, roughly dentated, foresaid, terminating in a beard or awn, which is rough to the touch, *fig.* 4. the inner valve lanceolate, flat, with a groove, truncated at top, and slightly emarginate, *fig.* 3. at the outer base of this valve awns a straight awn the length of the filaments, *fig.* 8.

NECTARY: two long-pointed, little Glumes, at the base of the germen, *fig.* 7.
STAMINA: three capillary Filaments, much shorter than the glumes of the corolla. Antherae small, of a yellowish green colour, *fig.* 5.
PISTILLUM: Germen ovate, downy. Styles two, reflexed, and villous, *fig.* 6.

Some of the grasses are noxious to the husbandman in one way, and useful in another. We have been informed, on the most respectable authority, that in the Isle of Thanet this grass is well known to the innkeepers, who call it Squirrel-tail Grass and God, that if horses feed on it for some time, the beards or awns of the fodder stick into their gums, and make them so sore, that they are in danger of being starved. The gentleman, who related to me this fact, informed me, that on the road he had a bill put into his hand, signifying, that at such an inn travellers ought depend on having good hay for their cattle, without any mixture of Squirrel-tail Grass.

It is chiefly on the edges of paths, at the bottoms of walls, and on the borders of fields, that we find this noxious grass: and in such situations it is extremely common in the neighbourhood of London. Fortunately it is seldom or never found in the body of pastures and meadows, and consequently it rarely occurs in our hay.

It continues to flower and produce seed during the greatest part of the summer.

We are careful to distinguish it from the *Hordeum pratense* of Mr. Hudson, which Linnaeus, contrary to the opinion of Ray, Vaillant, Halles, and other respectable Botanists, considers only as a variety of the present species.

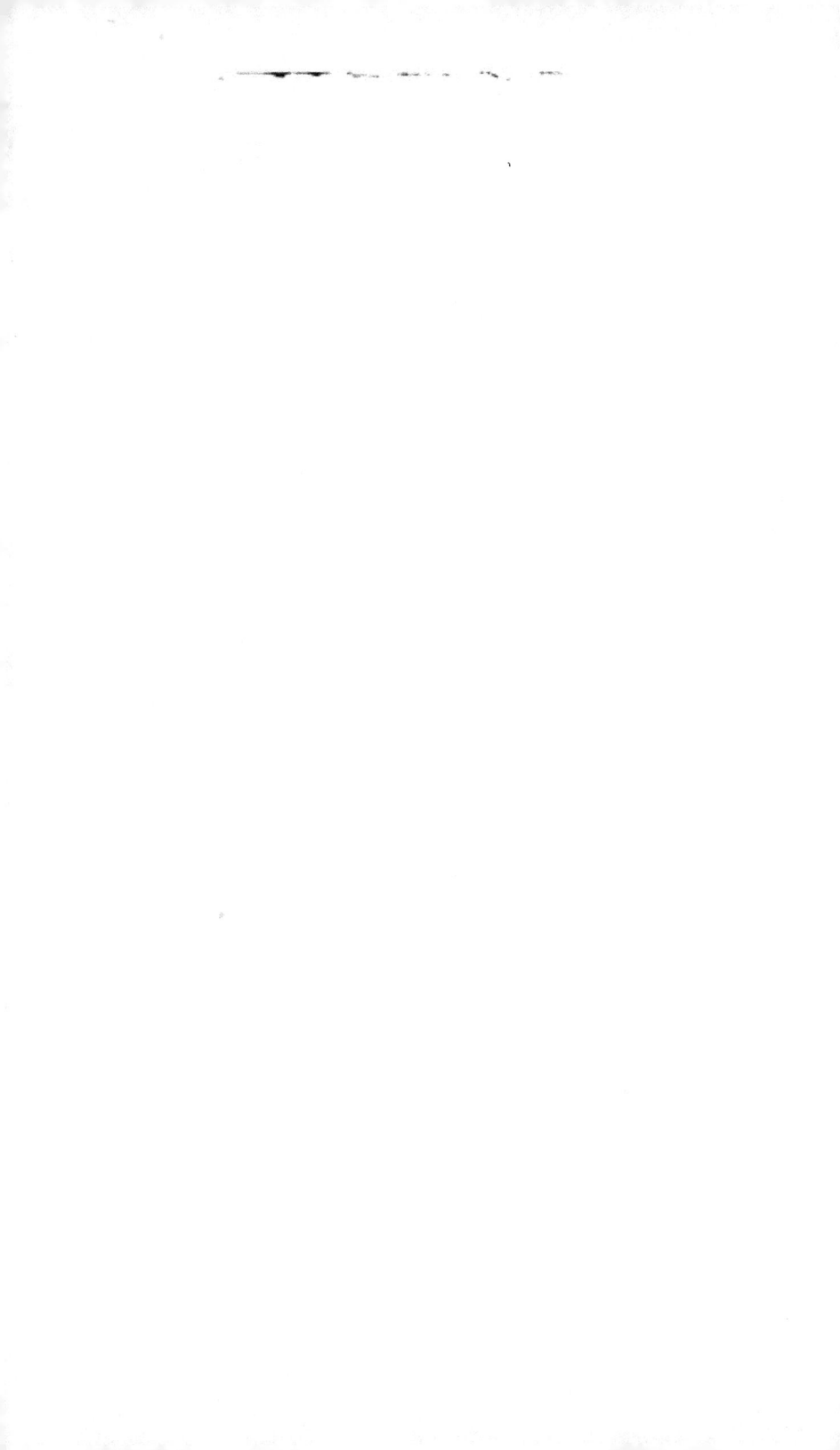

MELICA UNIFLORA. SINGLE-FLOWERED MELIC-GRASS.

MELICA *Lin. Gen. Pl.* TRIANDRIA DIGYNIA.

Cal. bivalvis, biflorus, rudimentum floris inter flosculos.

Raii Syn. Gen. 27. HERBÆ GRAMINIFOLIÆ FLORE IMPERFECTO CULMIFERÆ.

MELICA *uniflora* panicula rara, calycibus bifloris, flosculo altero hermaphrodito, altero neutro. *Retzii Fasc. Obf. Bot.* 1. p. 10 n. 9.

GRAMEN avenaceum locustis rarioribus. *Bauh. Pin.* p. 10.

GRAMEN avenaceum spica mutica rarior glumis. *Hift. Ox. III. t. 7. f.* 49.

GRAMEN avenaceum nemorense, glumis rarioribus ex fusco xerampelinis. *Raii Syn.* p. 405.

GRAMEN avenaceum rariore grano nemorose danicum. *Lob. Ad. P. Alt.* p. 465. *& I. B.* p. 434.

MELICA *anicus* petalis imberbibus, paniculis secundo nutante, glumis unifloris. *Hudfon. Fl. Angl. ed.* 2. p. 37. *Lightfoot Fl. Scot.* p. 95.

RADIX perennis, fibrofa.

CULMUS fimplex, fefquipedalis et ultra, foliofus, vaginis foliorum tegitur fubtriangulatus, fcaber, flriatus, ad bafin fordidè purpureus.

FOLIA culmo quinque rariùs, e fufco-viridia, plana, latiora unam cum dimidia vel duas fere lineas, in acutum mucronem finifim attenuata, fi digiti deorfum ducantur afpera, fuperne fubpilofis, marginibus ad lumbos acutiffimis traxilata, membrana breviffima, vix ulla, at quod valde fingulare, et natura dignum, folofum vento-communem, erectam, coloratam, ex interiore parte oris vaginæ oriuntur, notatur notabilem, uti cl. Retius obfervavit, *fig.* 8.

FLORES paniculati.

PANICULA rara, pedunculis inferioribus geminis altero breviore, trifloris, etiam feptem aut octo floris in hortis culta, fuperioribus folitaris.

SPICULÆ pedicellatæ, primo atro-purpureæ, muticæ, bifloræ.

CALYX: Gluma bivalvis, biflorus, coloratum, nitidus, valvula exteriore majore, ovata, concava, quinquenervi, fibnervacum, interiore minore, ovato-lanceolata, trinervi, *fig.* 1.

FLOS hermaphrod. feffilis, valvula exterior magna, ventricofa, marginibus interiorum amplectens, quæ planiufcula, marginibus membranaceis reflexis, præcipue prope bafin, *fig.* 2, 3.

fterilis pedunculatus, imperfectus, *fig.* 9. idem evolutus, *fig.* 10.

STAMINA: FILAMENTA tria, capillaria, brevia. ANTHERÆ florefcentes utrinque bifurcatæ, *fig.* 4.

PISTILLUM: GERMEN ovatum, glabrum, nitidum, flavefcens. STYLI duo bafi diferetis, divaricata. STIGMATA villofa, *fig.* 5.

NECTARIUM: Squamula minima, integra, ad bafin germinis, *fig.* 6.

SEMEN ovatum, nitidum, majufculum, nigricans, *fig.* 7.

ROOT perennial and fibrous.

STALK fimple, a foot and a half or more in height, leafy, where it is covered with the fheaths of the leaves fomewhat angular, rough and ftriated, of bottom of a dull purple colour.

LEAVES on the ftalk about five in number, of a yellowifh-green colour, flat, a line and a half or almoft two lines broad, terminating gradually in a point, rough if drawn backwards betwixt the fingers, on the upper fide fomewhat hairy, the edges of the leaves when magnified finely farrated, the membrane very fhort, fcarce any; but what is very remarkable and worthy notice, a fmall ovate leaf with a long point, upright, and coloured, rifes from the fore-part of the mouth of the fheath, till now undefcribed even by the celebrated *Retzius*, *fig.* 8.

FLOWERS growing in a panicle.

PANICLE loofe, the lowermoft flower-ftalks growing two together, the one fhorter than the other, bearing three flowers, and even feven or eight when cultivated in gardens, the uppermoft growing fingly.

SPICULÆ ftanding on little foot-ftalks, at firft of a dark purple colour, beardlefs, each containing two flowers.

CALYX: a Gluma of two valves, containing two flowers, coloured and fhining, the exterior of valve ovate, hollow, having five ribs, and terminated by a fhort point, the innermoft leaft, ovate-lanceolate, and three-ribbed, *fig.* 1.

FLOWER: the hermaphrodite one feffile, the outer valve large, bellying out, with its edges embracing the inner one, which is flattifh, the edges membranous and turned back, efpecially near the bafe, *fig.* 2, 3.

the *fterile* flower ftanding on a foot-ftalk, and imperfect, *fig.* 9. the fame unfolded, *fig.* 10.

STAMINA: three FILAMENTS, capillary and fhort. ANTHERÆ yellowifh and forked at each end, *fig.* 4.

PISTILLUM: GERMEN ovate, fmooth, fhining, and yellowifh. STYLES two, feparate at bottom and fpreading out. STIGMATA villous, *fig.* 5.

NECTARY: a very minute, entire fcale, at the bafe of the germen, *fig.* 6.

SEED ovate, fhining, rather large and blackifh, *fig.* 7.

This elegant fpecies, long fince noticed and defcribed by many of the old Botanifts, particularly by RAY, has been overlooked by LINNÆUS. Profeffor RETZIUS[*], in the firft fafciculus of his botanical obfervations, defcribes it anew, and gives it the name of *uniflora*, having found each fpicula to contain only one perfect flower. This name we therefore much readily adopt. Mr. Hudfon, in his *Flora Anglica*, has miftaken this plant for the *nutans* of LINNÆUS; and to the *nutans* has given the name of *montana*.

The delicacy and ftriking colour of its panicle, joined to its place of growth, readily diftinguifhes it from all our other graffes.

It grows plentifully in moft of the woods near London, and flowers in May and the beginning of June.

[*] *Andr. Joh. Retzii Fafciculus Obfervationum Botanicarum primus, cum figuris æneis. Lipfiæ,* 1779.

Melica ciliata

MELICA   *Lin. Gen. Pl.*   Triandria Digynia.

Cal. 2-valvis, 2-florus.   *Rudimentum floris inter flosculos.*

Raii Syn. Gen. 27.   Herba graminifolia flore imperfecto culmifera.

MELICA   cærulea panicula coarctata floribus cylindricis.   *Lin. Syst. Vegetab. p.* 113.

AIRA   cærulea foliis planis, panicula coarctata, floribus pedunculatis muticis convoluto fubulatis.   *Lin. Sp. Pl.* 95.   *Fl. Suec. n.* 67.

POA   spiculis fubulatis panicula rara coarcta.   *Fl. Lapp.* 39.

AIRA   cærulea.   *Scopoli Fl. n.* 91.

GRAMEN   arundinaceum enode minus fylvaticum.   *Bauh. Pin.* 7.   *Scheuch Agrost.* 209.

GRAMEN   pratense ferotinum, panicula longa purpurascente.   *Raii hist.* 1288.   *Morist. hist.* 3. *p.* 201. *f.* 8. *t.* 5. *f.* 22.

GRAMEN   pratense spica Lavendulæ.   *Merr. Pin.* 5.   *Raii Syn.* 404.   *Hudson. Fl. Angl. ed.* 2. *p.* 33.   *Lightfoot Fl. Scot. p.* 96.

RADIX perennis, fibrofa, fibris craffis, albidis feu fufcefcentibus, flexuofis, villofis.

CULMUS pedalis, fefquipedalis, aut bipedalis, bafi fub-bulbofus, erectus, unico tantum nodo, eoque prope bafin inftructo, fuperne nudus, lævis.

FOLIA plerumque tria, aut quatuor, palmaria, et ultra, ex cæruleo virefcentia, latiufcula, acuminata, rigidula, inferiora plana, fuperiora fubconvoluta, ad margines pilofa, Membrana nulla, Vagina brevis, ftriata.

FLORES paniculati.

PANICULA palmaris, et ultra, ramofa, ramis appreffis, hinc fubfpicata.

SPICULÆ bifloræ, trifloræ, et quadrifloræ, fæpius vero trifloræ, *fig.* 1, 2, 3. cum rudimento flofculi in plerifque, *fig.* 4, 5. juniores compreffæ, adultæ teretiufculæ, obtufæ, paululum divergentes.

CALYX bivalvis, valvulæ fubæquales, acutæ, carinatæ, ad margines purpureæ, *fig.* 6.

COROLLA bivalvis, valvulæ fubæquales, exterior majore, interiorem amplectente, trinervi, fubmucronata, ad margines purpurea, interiore binervi, pallidiore, obtufa, paulo breviore, *fig.* 7.

NECTARIUM: Squamulæ duæ, breviffimæ, latæ, truncatæ, emarginatæ, *fig.* 8.

STAMINA: Filamenta tria, capillaria; Antheræ bifurcæ, purpureæ, *fig.* 11.

PISTILLUM: Germen minimum, glabrum, fub-ovatum; Styli duo, ramofi, ad bafin ufque purpurei, *fig.* 9, 10.

ROOT perennial, fibrous, thick, whitish or brownish, crooked and vinous.

STALK a foot, a foot and a half, or two feet high, fomewhat bulbous at the bafe, upright, having only one knot, and that near the bafe, above naked and fmooth.

LEAVES for the moft part three or four, about a hand's-breadth in length, of a bluifh-green colour, rather broad, long-pointed, ftiffifh, the lower ones flat, the upper ones fomewhat rolled up, hairy at the edges, Membrane none, Sheath fhort and ftriated.

FLOWERS growing in a panicle.

PANICLE a hand's-breadth or more in length, branched, the branches clofing together fo as to form a kind of fpike.

SPICULÆ containing two, three, and four flowers, but moft commonly three, *fig.* 1, 2, 3. with a rudiment of a flower in moft of them, *fig.* 4, 5. the young ones flattened, the full-grown ones roundifh, obtufe, flightly diverging.

CALYX compofed of two valves, the valves nearly equal, pointed, keeled, the edges purple, *fig.* 6.

COROLLA compofed of two valves, the valves nearly equal, the outer one, which is largeft, embracing the inner one, three-ribb'd, flightly pointed, the edges purple, the inner valve two-ribb'd, paler, obtufe, and a little fhorter, *fig.* 7.

NECTARY: two very fhort, broad, truncated, emarginate Scales, *fig.* 8.

STAMINA: three capillary Filaments; Antheræ forked at each end, and purple, *fig.* 11.

PISTILLUM: Germen very minute, fmooth, and fomewhat ovate; Styles two, branched down to the bottom, and purple, *fig.* 9, 10.

Our readers, on perufing the above defcription, will quickly perceive, that this grafs does not accord, in every refpect, with the characters of a *Melica*; it has, in general, too many flowers: yet, as the effential part, the rudimentum flofculi, is found in moft of the fpiculæ, it cannot, perhaps, be more judicioufly arranged.

Linnæus, at different periods, appears to have entertained a different opinion of it: in his *Flora Lapponica*, he confiders it as a *Poa*; in his *Species Plantarum* and *Flora Suecica*, as an *Aira*; and, laftly, in his *Syftema Vegetabilium*, makes it a *Melica*.

If the Species be examined when the plant is young, they are certainly very Poa-like, being pointed, flattened, and containing ufually from three to five flowers; as they advance, their form alters, they become rounded, and more like the flowers of the *Aira aquatica*: if the rudimentum flofculi were wanting, it would be difficult to fay with which of the two genera it fhould be placed; that being prefent, the difficulty vanifhes, and we clafs it at once with the Melica.

Two ftriking peculiarities diftinguifh this grafs: the ftalk has only one knot, and that near its bafe; and not only its flamina, but its ftigmata alfo, are of a deep purple colour.

Merret's name of Gramen Spica Lavendulæ, is very expreffive of its appearance when in flower.

It is a very common grafs on wet moors and heaths, and flowers from July to the end of September; it is harfh and bitter, and therefore does not form at all adapted to agricultural purpofes; it varies greatly in fize.

Mr. Lightfoot, in his *Flora Scotica*, informs us, that in the Ifle of Skie, the fifhermen make ropes for their nets of this grafs, which they find by experience will bear the water well without rotting. Scheuchzer fays, that befoms are fometimes made of the ftraws.

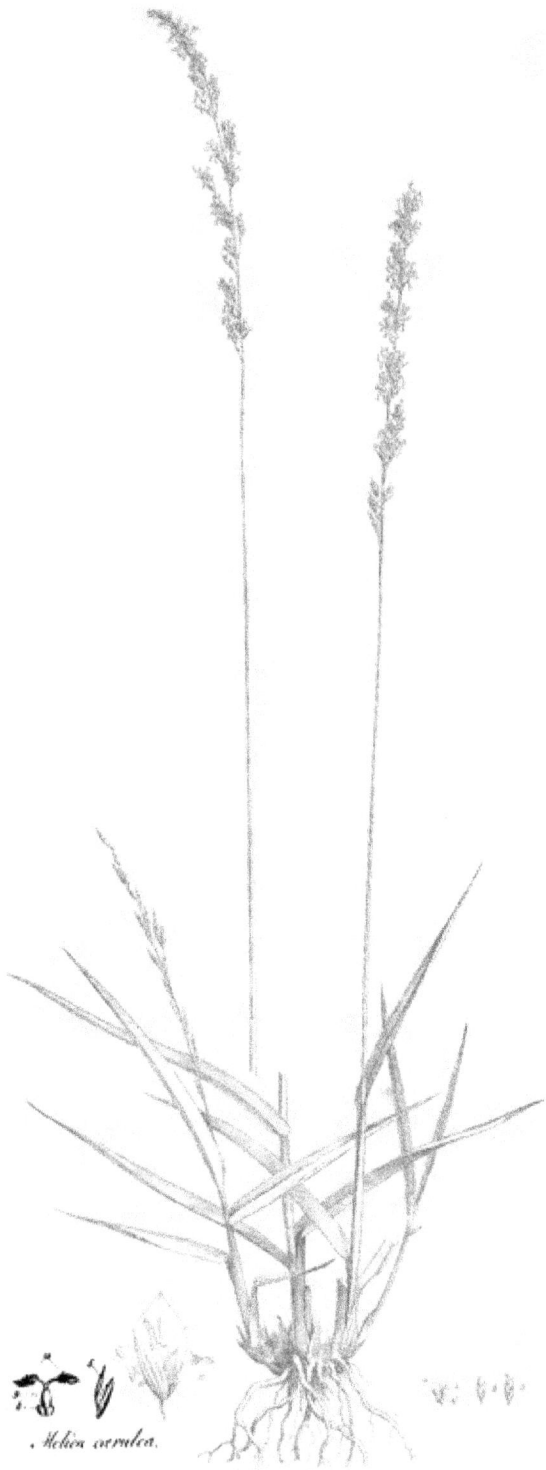

Molinia cœrulea.

POA AQUATICA.   WATER MEADOW GRASS.

POA   *Lin. Gen. Pl.*   TRIANDRIA DIGYNIA.

    Cal. 2-valvis, multiflorus. *Spicula* ovata : valvulis margine scariosis acutiusculis.

Raii Syn. Gen. 27.   HERBÆ GRAMINIFOLIÆ FLORE IMPERFECTO CULMIFERÆ.

POA   *aquatica* panicula diffusa, spiculis sexfloris linearibus.   *Lin. Syst. Vegetab. p.* 97.   *Sp. Pl. p.* 98.   *Fl. Suec. n.* 26.

POA   altissima, foliis latissimis, panicula amplissima, locustis distichis multifloris.   *Haller hist. n.* 1454.

POA   *aquatica.*   *Scopoli Fl. Carn. n.* 105.

GRAMEN   aquaticum paniculatum latifolium, *Bauh. Pin.* 3.

GRAMEN   aquaticum majus. *Ger. emac.* 6.   *Raii Syn. p.* 411.   Great Water-Reed-Grass. *Hudson. Fl. Angl. ed.* 2. *p.* 38.

RADIX perennis, repens.

CULMUS tripedalis, ad sexpedalem, erectus, foliosus, crassitie culmi arundinacei, superne ubi nudus, teres, lævis, subtilissime striatus ; geniculis flavescentibus.

FOLIA sesquiuncem aut unciam fere lata, utrinque glabra, tenuissime striata, carinata, carina marginibusque asperis, ad basin folii utrinque macula triangularis flava, vagina glabra, striata, carina prominente. membrana brevis obtusa.

PANICULA maxima, semipedalis, aut pedalis, erecta, ramosissima.

PEDUNCULI subtriquetri, scabri, superne flexuosi.

SPICULÆ lanceolatæ, subcompressæ 6—8. floræ, colore ex spadiceo et viridi misto.

CALYX: Gluma bivalvis, valvulæ membranaceæ, uninerviæ, ovatæ, concavæ, interiore breviore et acutiore.

COROLLA bivalvis, valvulæ subæquales, obtusæ, exteriore majore, concava, nervosa, ad basin tuberculata, interiore planiuscula.

STAMINA: FILAMENTA tria, alba, capillaria, ANTHERÆ oblongæ, utrinque bifidæ, flavæ aut purpureæ.

PISTILLUM: GERMEN ovatum, glabrum; STYLI duo, superne ramosi, inferne nudi, paulo infra apicem prodeuntes.

NECTARIUM: squamula parva truncata ad basin germinis.

SEMEN tectum, hinc convexum, striatum, inde concavum, pallide fulcum.

ROOT perennial, and creeping.

STALK from three to six feet high, upright, leafy, the thickness of a reed straw, on the upper part where it is naked, round, smooth, very finely grooved; the joints yellowish.

LEAVES half an inch and almost an inch broad, smooth on both sides, very finely grooved, keeled, the keel as well as the edges rough, the base of the leaf on each side is marked with a yellow triangular spot, the sheath is smooth and striated, the keel prominent, the membrane short and obtuse.

PANICLE very large, from six inches to a foot in length, upright, very much branched.

FLOWER-STALKS somewhat three-cornered, rough, crooked above.

SPICULÆ lanceolate, somewhat flattened, containing from six to eight flowers, variegated with green and purple.

CALYX: a Gluma of two valves, the valves membranous, one-ribbed, ovate, concave, the innermost shorter and more pointed than the other.

COROLLA composed of two valves, which are nearly equal, obtuse, the outer one largest, concave, ribbed, with a small tubercle at the base, the inner one nearly flat.

STAMINA: three, white, capillary FILAMENTS; ANTHERS oblong, bifid at each end, yellow or purple.

PISTILLUM: GERMEN, ovate, smooth; STYLES two, branched above, naked below, proceeding from a little below the top.

NECTARY: a small truncated scale at the base of the germen.

SEED covered, convex and striated on one side, concave on the other, of a pale brown colour.

The *Poa aquatica* is one of the largest as well as the most useful of our grasses ; it constitutes a great part of the riches of Cambridgeshire, Lincolnshire, and other counties, where draining the land by means of windmills has taken place, inasmuch tracts of territory that used to be overflown and produce useless aquatics, but which still retain much moisture, are, by the above process, spontaneously covered with this grass, which not only affords rich pasturage for their cattle in the summer, but forms the chief part of their winter fodder.

It has a powerfully creeping root, and bears frequent mowing well (we have known it cut thrice in one season in the vicinity of the Thames) ; hence it is apt to gain the ascendancy over, rather than be overcome by other plants.

It grows not only in very moist ground, but in the water itself: like the Cats-tails, Bure-reed, and several other plants of that kind, it soon fills up the watery ditches which surround the meadows in which it grows, and occasions them to require frequent cleaning ; in this respect it is a formidable plant, even in slow rivers.

In the Isle of Ely, they have a particular method of cleansing the rivers, which are liable to be soon choked up by the Arrow-head, Water-lilies, Reeds, &c. by means of an instrument called a Bear, which is an iron roller, to which a number of pieces of iron, like small spikes, are fixed ; this is drawn up and down the river by horses, which travel on the banks, and tearing up every plant by the roots, they float and are carried away by the stream.

The *Poa aquatica* not only affords sustenance to cattle, but is a favourite food of the Caterpillar of the Gold-spot Moth *(Phalæna Festucæ, Lin.)* which LINNÆUS describes as feeding on the *Festuca fluitans,* but which feeds with us chiefly on this grass : the Moth proceeding from this larva, is one of the most beautiful which this country produces: the Caterpillar being smooth and of a green colour, is not easily distinguished from the grass on which it feeds ; when full grown, it usually bends down the top of one of the leaves, and underneath it, makes a thin spinning, in which it changes to chrysalis ; this spinning, from its whiteness, is easily discovered ; but we small apprise our readers, that these Caterpillars are not very numerous, and that they will be fortunate if they find one or two after a long search ; the Moth, Caterpillar, and Chrysalis, are figured in ALBIN's English Insects ; but a much better painting of the Moth may be seen in ROESEL, Tom. 3, Tab. 50. We have generally found them at the commencement of harvest, when the wheat has been in sheaf ; the Moth comes forth in a week or two.

We observed in the Isle of Ely, a much larger Caterpillar, when full-grown, nearly the size of the *Ph. Pinastria,* hairy and very beautiful, not uncommon on this grass ; but not having the proper convenience for breeding it, we are as yet unacquainted with the Moth it produces, but suspect it will prove a non-descript.

The *Poa aquatica* flowers as late as August and September.

Poa aquatica

SHERARDIA ARVENSIS. FIELD SHERARDIA.

•

SHERARDIA *Lin. Gen. Pl.* TETRANDRIA MONOGYNIA.

Cor. 1-petala, infundibuliformis. *Semina* 2, tridentata.

Raii Syn. Gen. 13. HERBÆ STELLATÆ.

SHERARDIA *arvensis* foliis omnibus verticillatis, floribus terminalibus. *Lin. Syst. Vegetab. p.* 135. *Spec. Pl. p.* 149. *Fl. Suec. n.* 120.

SHERARDIA foliis senis lanceolatis, floribus sessilibus umbellatis. *Haller. Hist. n.* 734.

SCHERARDIA *arvensis. Scopoli Fl. Carn. n.* 143.

RUBEOLA arvensis repens cærulea. *Bauh Pin.* 334.

RUBIA minor repens cærulea. *Parkins. p.* 276.

RUBEOLA parvo flore cæruleo se spargens. *J. B. III.* 719. *Raii Syn. p.* 225. Little field Madder. *Hudson Fl. Angl. ed.* 2. *p.* 66. *Lightfoot Fl. Scot. p.* 114.

RADIX annua, fibrosissima, fibrillis rufis.

ROOT annual, extremely fibrous, the small fibres reddish brown.

CAULES primares, spithamaei et ultra, procidui, asperi, tetragoni.

STALKS a hand's breadth, half a foot or more in length, laying on the ground, rough and four-cornered.

FOLIA superiora verticillata, seni, seu quini, inferia lanceolata, inferiora numero seorsim decrescent, et latiora fiunt, infima saepius terna, ovata, semiverticillata, omnibus mucronatis, superne scabra.

LEAVES: those on the upper part of the stalk growing in whirls, five or six together, the leaves lanceolate, the lower leaves gradually decreasing in number, and becoming broader, the lowermost generally growing three together, ovate, and forming half a whirl, all of them terminating in a short point, and rough on the upper side.

FLORES umbellati, sessiles, parvi, lacte purpurei.

FLOWERS growing in umbels, sessile, small, of a bright purple colour.

PEDUNCULI axillares, solitarii, tetragoni, peracta florescentia longitudine foliolorum.

FLOWER-STALKS growing from the alae of the leaves, solitary, four-cornered, when the flowering is over the length of the leaves.

CALYX Involucorum octophyllus, foliolis lanceolatis, carinatis, ciliatis.

CALYX: an Involucrum of eight leaves, which are lanceolate, keeled and edged with hairs.

CALYX Perianthium parvum, 6-dentatum, superum, peruldens, *fig.* 1.

CALYX: a small Perianthium, having six teeth, placed on the top of the germen and permanent, *fig.* 1.

COROLLA monopetala, infundibuliformis. *Tubus* cylindraceus, longus. *Limbus* quadripartitus, planus, laciniis acutis, *fig.* 2.

COROLLA monopetalous, funnel-shaped. *Tube* cylindrical and long. *Limb* flat, divided into four sharp segments, *fig.* 2.

STAMINA: FILAMENTA quatuor ad apicem tubi posita, demisso pollice reflexa. ANTHERAE simplices, pallide purpureae, *fig.* 3.

STAMINA: four FILAMENTS placed at the top of the tube, turning back on the shedding of the pollen. ANTHERAE simple, pale purple, *fig.* 3.

PISTILLUM: GERMEN solymum, oblongum, inferum, *fig.* 4. STYLUS filiformis, superne bifidus. STIGMATA capitata, *fig.* 5.

PISTILLUM: GERMEN double, oblong, beneath the calyx, *fig.* 4. STYLE filiform, bifid at top. STIGMATA forming two small heads, *fig.* 5.

PERICARPIUM nullum: fructus oblongus, coronatus, longitudinaliter in duo semina separabilis.

SEED-VESSEL none; the fruit oblong, crowned, separable longitudinally into two seeds.

SEMINA bina, oblonga, apice tribus acuminibus notata, hinc convexa inde plana, *fig.* 6, 7.

SEEDS two together, oblong, furnished at top with three points, convex on one side and flat on the other, *fig.* 6, 7.

TOURNEFORT considered this plant as a species of *Aparine*. The more accurate DILLENIUS made a new genus of it, to which he gave the name of his friend and patron, that excellent English Botanist Dr. SHERARD. *Vid. Dill. Nov. Pl. Gen. p.* 96.

This small annual is a native of our corn fields, and common almost every where, flowering during the greatest part of the summer.

There is a neatness in its blossoms almost sufficient to recommend it as an ornamental plant: to any other use it does not appear to have the least pretensions.

Sherardia arvensis

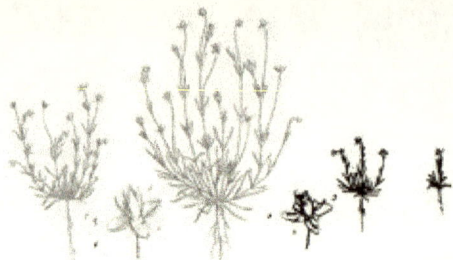

Sagina apetala

Sagina apetala. Annual Pearl-wort.

SAGINA *Lin. Gen. Pl.* Tetrandria Tetragynia.

Cal. 4-phyllus. Petala 4. Caps. ovularis, 4-valvis, polysperma.

Rail Syn. Gen. 24. Herba pentapetala vasculifera.

SAGINA *apetala nobis notata, caule erectiusculo pubescente.*

SAGINA *apetala caule erectiusculo pubescente, floribus alternis* apetalis. *Lin. Mantiss.* 559. 598. *Vegetab.* p. 162.

SAGINA *erectior erecta, ramis annuis, floribus apetalis. Ard Spec.* t. 9. 22. t. 8. fig. 1.

SAXIFRAGA Anglica Alsinella annua. *D. Pet. Hist. Nat. Off.* t. 6. § 9. t. 9. t. 7. *Rail Syn.* p. 345. Annual Pearl-wort.

ALSINE Saxifraga graminifolia, foliolis tetrapetalis herniolis et muscosis. *Pluk. Alm.* t. 74. f. 2.

SAGINA *procumbens var. ß. Hudson Fl. Angl. ed.* 2. p. 73.

RADIX annua, fibrosa.	ROOT annual and fibrous.
CAULES plures, primo procumbentes, dein erecti, teretes, filiformes, aliqui, teretes, filiformes, hispiduli, modeli.	STALKS several, at first procumbent, afterwards upright, from one to three inches or more in height, round, filiform, somewhat hispid, and jointed.
FOLIA opposita, lineari-subulata, brevia, mucronata, hispidula.	LEAVES opposite, linear, and somewhat awl-shaped, short, terminated by a fine point, and somewhat hispid.
FLORES alterni, pedunculati.	FLOWERS alternate, and standing on foot-stalks.
PEDUNCULI apice primo curvaturo, demum erecti, pilis rara vestiti.	FLOWER-STALKS first drooping at top, finally upright, covered with a few hairs.
CALYX: Perianthium tetraphyllum subindo pentaphyllum, foliolis ovatis, obtusis, concavis, brevibus, permanentibus, marginibus purpureo scariosis, fig. 1.	CALYX: a Perianthium of four, sometimes five, ovate, obtuse, hollow, smooth, permanent leaves, with purplish edges, fig. 1.
COROLLA: Petala plerumque quatuor, minutissima, nulla occulo vix conspicua, alba, obcordata, fig. 2.	COROLLA: generally composed of four Petals, which are extremely small, and scarcely visible to the naked eye, white and inversely heart-shaped, fig. 2.
STAMINA: Filamenta quatuor alba, calyce breviora, Antheræ albæ, fig. 3.	STAMINA: four white Filaments, shorter than the calyx, Antheræ white, fig. 3.
PISTILLUM et Capsula ut in Sagina procumbente.	PISTILLUM and Capsule as in the procumbent Pearl-wort.

Mr. Ray, in his Synopsis, considers this species as distinct from the procumbent; and informs us, that it differs from it not only in the colour of its stalks and leaves, which are of a browner hue, but that it has an annual root; and that it does not put forth roots at the joints as the procumbent does, he refers to a figure given of it by Plot in his Natural History of Oxfordshire.

Notwithstanding Ray's description, and Plot's figure, Linnæus, in his *Spec. Plant.* considered it only as a variety of the procumbent; but afterwards, more fully convinced by the description and figure given of this plant by Arduinus, an Italian Botanist, he adopts it in his second *Mantissa* as a species. It appears, by Mr. Hudson's quotations, that he has been no stranger to the observations of these authors; but, in opposition to them all, he continues it only as a variety.

From a thorough conviction of the propriety of Mr. Ray's conduct in making it a species, we have given a figure of it, and shall not only confirm his account, but give a few additional remarks of our own, which we presume may finally settle this matter.

The distinction of an annual and perennial root, though it cannot be admitted, perhaps, in all cases as a specific character, must be allowed to have considerable weight. To ascertain the constancy of this character we have for several years cultivated the two plants close together, on a soil with peculiar care; the result has been that the apetala has proved as regular an annual as the Draba verna, while the procumbent has continued green through the winter; and we have no doubt but this always is the case with these plants, when they grow in their natural situations.

The procumbent is always procumbent; and when it grows, as it most commonly does, in moist situations, it roots and spreads on the ground. The stalks of the apetala, when the plant is young, spread on the ground, but as it advances to maturity they rise up, and, if several grow together, become quite erect. Where the plants grow singly, and in a dry situation, they neither acquire the same height, nor the same degree of uprightness. Sometimes this species is found on moist dusty walls, much taller and more branched than the specimens we have figured; but whether the plants of the apetala be small or large, their stalks and leaves are always hairy, while in the procumbent they are perfectly smooth, the hairs are visible to the naked eye, and when magnified have so little globule in their extremities, as those of the Apetala fagellata have, which comes very near in its appearance to the Pearl-wort; thus we find these three distinct plants may, with certainty, be distinguished by this little circumstance.

The apetala is a smaller plant than the procumbent, and much more in its stalks. Its leaves are also shorter by almost one-half, and less succulent; and these, as far as we have observed, are the chief differences.

From its name one would be led to suppose, that it was perfectly apetalous; and both Linnæus and Arduinus describe it as such. We have generally found it with petals, but so minute, indeed, as almost to require a magnifier to render them visible. These petals we have given a magnified view of, and have represented the plant in the several states in which it is found in dry situations.

Mr. Ray does not appear to have had an idea of its being a common plant, as he mentions the particular spots where it was to be found; with us there is no plant more abundant, especially on walls, in gravel walks, where it is a troublesome weed, and on barren heaths.

It flowers in May and June. There is, perhaps, scarce any plant that is quicker in ripening its seeds.

In our examination of this plant we found the egg of a very small moth glued to an unripe capsule, the seeds of which were probably destined to feed its caterpillar.

POTAMOGETON CRISPUM. CURLED PONDWEED, or
GREATER WATER CALTROPS.

POTAMOGETON *Lin. Gen. Pl.* TETRANDRIA TETRAGYNIA.

Cal. o. *Petala* 4. *Stylus* o. *Sem.* 4.

Raii Syn. Gen. 5. HERBÆ FLORE IMPERFECTO SEU STAMINEO VEL APETALO POTIUS.

POTAMOGETON *crispum* foliis lanceolatis alternis oppositisve undulatis ferratis. *Lin. Syst. Vegetab.* p. 141. *Sp. Pl.* p. 183. *Fl. Suec. n.* 148.

POTAMOGETON. *Hall. Hist. n.* 848.

POTAMOGETON *crispum. Scopoli Fl. Carn. n.* 181.

POTAMOGETON foliis crispis seu lactucæ ranarum, *Bauh.* p. 465.

POTAMOGETON seu fontinalis crispa. *I. B. III.* p. 778.

TRIBULUS aquaticus minor Quercus floribus. *Ger. em.* 1282.

TRIBULUS aquaticus minor prior. *Park.* 1248. *Raii Syn.* p. 149. The greater Water Caltrops. *Hudson Fl. Angl.* p. 75. *Lightfoot Fl. Scot.* p. 122.

RADIX perennis, repens.
CAULES plurimi, variæ longitudinis, fordidè carnei, subdiaphani, compressi, utrinque sulcati, ramosi.

VAGINÆ breves, concolores, vix distinguendæ.

FOLIA sessilia, lanceolata, obtusa, subdiaphana, crispa, lucentia, acrida, nitorem, serrulata, inferiora luce alterna, superioribus oppositis.

PEDUNCULI axillares, bi seu triunciales, crassiusculi, subcompressi.

FLORES spicati, sex sive octo, sessiles.
CALYX nullus.
COROLLA: PETALA quatuor, subrotunda, obtusa, concava, argenteolæ, primum erecta, dein paulatim, dividua, e sordido viridia, *fig.* 1.

STAMINA: FILAMENTA quatuor, brevissima, vix distinguenda. ANTHERÆ breves, dilymæ, albæ, *fig.* 2.
PISTILLUM: GERMINA quatuor, ovato-acuminata. STYLES nulli. STIGMATA obtusa, *fig.* 3.
SEMINA quatuor, nuda, majuscula, sordide virentia, utrinque compressa, extorne ad basin denticulata, *fig.* 4.

ROOT perennial and creeping.
STALKS numerous, of various lengths, of a dirty flesh-colour, somewhat transparent, flattened, with a groove on each side, and branched.

SHEATH: short, of the same colour as the stalks, scarcely to be distinguished.

LEAVES sessile, lanceolate, obtuse, somewhat transparent, curled, somewhat to the touch, shining, three-ribbed, sharply and finely serrated, the lower ones alternate, the upper ones opposite.

GENERAL FLOWER-STALKS growing from the axe of the leaves, two or three inches in length, thickish, and somewhat flattened.

FLOWERS six or eight, growing in a spike, and sessile.
CALYX wanting.
COROLLA: four PETALA roundish, obtuse, hollow, concave, of a little claw, at first upright, afterwards spreading and deciduous, of a greenish brown colour, *fig.* 1.

STAMINA: four FILAMENTS, very short, scarcely to be distinguished. ANTHERÆ short, having two separate lobes, of a white colour, *fig.* 2.
PISTILLUM: GERMINA four, ovate, with a long point. STYLES none. STIGMATA obtuse, *fig.* 3.
SEEDS four, naked, rather large, of a dirty green, flattened on each side, toothed externally at the base, *fig.* 4.

Most of the plants of this genus have creeping roots, which, penetrating easily through the mud, cause them to spread very fast, so as soon to fill up a pond or slow river, if unmolested.

We have observed, that ducks very readily eat not only the seeds, but the leaves of the present species, which is one of the most common. The introduction of water-fowl may therefore probably prevent this species at least, and perhaps some of the others, from increasing too much.

It flowers in June and July.

Potamogeton crispum

Atropa Belladonna.

ATROPA BELLADONNA. DWALE, OR DEADLY NIGHTSHADE.

ATROPA *Lin. Gen. Pl.* PENTANDRIA MONOGYNIA.

Cor. campanulata. *Stam.* distantia. *Bacca* globosa, 2-locularis.

Raii Syn. Gen. 16. *Herbæ Bacciferæ.*

ATROPA *Belladonna* caule herbaceo, foliis ovatis integris. *Lin. Syst. Vegetab. ed.* 14. *p.* 221.
Sp. Plant. p. 260.

BELLADONNA caule herbaceo, brachiato, foliis ovato lanceolatis, integerrimis. *Haller. hist.* n. 579.

BELLADONNA *trichotoma. Scopoli Fl. Carn.* n. 255.

SOLANUM melanocerasus. *Bauh. pin.* 166.

SOLANUM lethale. *Ger. emac.* 340. *Parkins.* 346. *Raii Syn. p.* 265. Deadly Nightshade, Dwale. *Hudson Fl. Angl. p.* 93. *Lightfoot Fl. Scot. p.* 144. *Jacquin Fl. Austr. t.* 309.

RADIX perennis, crassa, albida, ramosa, repens.

ROOT perennial, thick, whitish, branched, and creeping.

CAULES plures, basi digitum crassi, tripedales et ultra, erecti, herbacei, teretes, ramosi, in apricis sordide purpurei, pubescentes.

STALKS several, at bottom the thickness of one's finger, three feet or more high, upright, herbaceous, round, branched, in exposed situations of a dingy purple colour, downy.

FOLIA petiolata, ovata, acuta, integerrima, utrinque lævia, venosa, ad latera caulis ramorumque gemina et magnitudine inæqualia, inter quæ pedunculus uniflorus et sæpius solitarius egreditur.

LEAVES standing on footstalks, ovate, pointed, perfectly entire, smooth on both sides, veiny, growing in pairs (but unequal in size) from the sides of the stalks, from betwixt them rises the flower-stalk supporting one flower, and usually single.

PEDUNCULI teretes, viscidi, ad flores paululum incrassati.

FLOWER-STALKS round, viscid, thickened somewhat next the flowers.

FLORES cernui, inodori, sordide purpurei, subviscidi, externe nitidi, venosi.

FLOWERS drooping, scentless, of a dingy purple colour, somewhat viscid, externally glossy and veiny.

CALYX: PERIANTHIUM monophyllum, quinquepartitum, angulatum, laciniis ovato-acuminatis, inæqualibus, viscidis, *fig.* 1.

CALYX: a PERIANTHIUM of one leaf, deeply divided into five segments, angular, the segments ovate-acuminate, unequal, and viscous, *fig.* 1.

COROLLA monopetala, campanulata. *Tubus* brevissimus, albus, subpentagonus; *Limbus* ventricosus, ovatus, ore quinquesido, patulo, laciniis subæqualibus, *fig.* 2.

COROLLA monopetalous, bell-shaped; *Tube* very short, white, slightly five-cornered; *Limb* bellying out, ovate, mouth spreading, divided into five equal segments, *fig.* 2.

STAMINA: FILAMENTA quinque, albida, quorum duo paulo breviora, inferne paulo crassiora, pilosa, apice incurva, hoc pondine tubuli; ANTHERÆ magnæ, didymæ, lutescentes, remotæ, *fig.* 3.

STAMINA: five FILAMENTS, whitish, two of which are a little shorter than the rest, somewhat thicker towards the base, and hairy, bent down at top, the length of the tube; ANTHERÆ large, double, yellowish, and remote, *fig.* 3.

PISTILLUM: GERMEN semiovatum, utrinque sulcatum, ad basin glandula lutescente cinctum; STYLUS filiformis, staminibus longior, inclinatus; STIGMA capitatum, assurgens, transverso-oblongum, bilabiatum, viride, *fig.* 4.

PISTILLUM: GERMEN semiovate, with a groove on each side, surrounded at bottom with a yellowish gland; STYLE thread-shaped, longer than the stamina, inclined downwards; STIGMA forming a little head, transversely oblong, two-lipp'd, of a green colour, *fig.* 4.

PERICARPIUM: BACCA atra, nitida, subrotunda, saporis dulcis, bilocularis, *fig.* 5. 6.

SEED-VESSEL: a black, glossy, roundish BERRY, of a sweet taste, with two cavities, *fig.* 5. 6.

SEMINA plurima, fusca, irregularia, *fig.* 7.

SEEDS numerous, brown, and irregular in shape, *fig.* 7.

Obs. Semina fuscescunt priusquam Bacca nigrescit.

Obs. The seeds turn brown before the Berry becomes black.

The rage for building, joined to the numerous alterations perpetually making in the environs of London, have been the means of extirpating many plants which formerly grew plentifully around us. To this cause we are to attribute the loss of the present plant, which the late Sir WILLIAM WATSON and Mr. STANESBY ALCHORNE of the Tower, gentlemen eminent for their knowledge of British plants, have often allured me grow, within their remembrance, in several places near town; happily we are now under the necessity of going much farther into the country, if we wish to see it grow wild. We have frequently noticed it in many of the chalk-pits in Kent, and in both shady and exposed situations elsewhere: in particular, we remember to have seen it growing in great abundance on Keep-Hill, near High Wycomb, Buckinghamshire. Close by the spot where we observed it, there chanced to be a little boy; I asked him, if he knew the plant? He answered "Yes, it was nought man's cherries." I then inquired of him, if he had ever eaten any of the berries? He said he had, with several other children from an adjoining poor-house, and that it made them all very sick, but that none of them had died.

Was not this plant studiously destroyed wherever it is found wild, it would be much more common than it is; for there are few plants to which nature has been so liberal in the means of increase: it has a very large perennial root, which runs deep into the earth, multiplies greatly, and frequently creeps under ground to a great distance; added to this, its berries are very numerous, and contain a prodigious quantity of seeds.

Forbidding

Forbidding as this plant may appear to some, its large glossy berries are certainly a great temptation to children; and, therefore, gentlemen, if they have the plant in their gardens, should never suffer it to ripen its fruit.

It flowers in June and July; its berries are ripe in August and September.

Numerous instances of the pernicious, and even deleterious effects of the deadly Nightshade are on record; among others, such of our readers as are fond of history will not be displeased with the prolixity of the following account taken from *Blair's Pharmaco-Botanologia, p. 81.*

" The *Solanum Lethale* seems to produce the same effects with the *Hyoscyamus, Cynoglossum,* and other " intense Narcoticks, which usually, before they affect the person with sleep, produce delirious and maniacal " symptoms; however it is an herb of so pernicious a nature, that scarce any Author who treats of it fails, " from proper observation, or good information, to give dismal instances of its bad effects. *Simon Pauli* " refers us to *Lobelius* his *Adversaria,* and *Rodens à Stapel.* Mr. *Ray's* accounts of what happened to a " Mendicant Friar, upon the taking a glass of the infusion of it in mellow wine, gives a good account of the " various symptoms it produces. In a short time, he became delirious, after a little (*Cachinnus*) a grinning laughter " like the *Risus Sardonicus* succeeded; after that several irregular motions; and at last a real madness, and " such a stupidity at those that are foolishly drunk have; which after all was cured by a draught of vinegar. " Mr. *Miller* mentions several Children at *Croydon,* who not long since were poisoned. Another instance " of its bad effects has fallen under my own observation: two or three persons not far from hence, having " got into a gentleman's garden, were delighted with the black berries of the *Solanum Lethale,* and eat some of " them; it was very pleasant (within a short time after) to see their frantic humours, gestures, and speeches; " but upon their taking of emeticks in due time, they were cured. It is worthy of recital what Mr. *Ray* " tells us happened to a *Lady of Quality* of his acquaintance, who having a small ulcer a little below her " eye, which she suspected to be cancrous, she applied a bit of the leaf of this *Solanum,* which so relaxed " the *Tunica Uvea* in one night, that she could not contract the *Pupilla* the next day; so that the *Pupilla* of " the one eye was four times as big as the other; and upon the removal of the leaf, the fibres recovered their " muscular tone by degrees: and, lest this should seem to be merely accidental, she repeated the experiment " three times, at which Mr. *Ray* himself was present.

" But the most memorable instance of the direful effects of this *Plant* is to be seen recorded by the cele- " brated *Buchanan,* in his History of Scotland, by which we may observe how the Almighty God can " convert the most deadly poisons into the finest antidotes, for those whom he has a mind to preserve. This " obliges me to make a digression, not altogether unsuitable, since it gives the botanical description of a " *Plant,* writ about a hundred and fifty years ago, by one who himself was no professed Botanist, the use " made of it, and the wonderful effects it produced.

" In the reign of *Duncan* I. *King* of Scotland (who was afterwards murdered by *Macbeth* the *Tyrant*) " *Harold the Dane* invaded *England,* not long before the days of *King William the Conqueror: Sueno,* his " brother, at the same time invaded *Scotland.* Upon his landing in *Fife,* he obtained a signal victory, which " obliged the *King of Scotland,* with the remainder of his routed forces, to retire to *Bertha* (an ancient town " of great note situated on the river *Tay,* which was not long after destroyed by an inundation, and out " of whose ruins the town of *Perth* was built, and now stands upon the same river, two miles nearer the " sea) and pursued them so closely, that he laid siege to the town both by land and water. The *Scots* were " put to great straits, not for want of provisions, but for want of arm to repel the besiegers. *King Duncan* " was a peaceable unactive man; he had sometime before committed the government to the management of " *Bancho,* of a cunning and subtle wit; and to *Mackbeth,* of a fierce, bold, aspiring spirit. *Mackbeth* " went to the country to raise a reinforcement, while *Bancho* treated with the enemy, and first obtained a " cessation of arms, and then spun out time by framing of articles of peace. The *Danes* wanted provisions, " but abounded with men; the *Scots* abounded in provisions, but wanted men. The truce was equally " acceptable to both, especially to the *Danes,* who for the present expected plenty of all things, and for the " future the conquest of a whole kingdom. Care was immediately taken by the *Scots* to afford them all " manner of liquors, both wine and ale, and they continued to mix with them a good quantity of the " Deadly Nightshade (this *Solanum Lethale,* or *Somniferum*) of which we now treat. The bait took; the " *Danes* drank plentifully, and were all intoxicated; mad with this poisonous juice, and asleep through " drunkenness, the *Scots* fell upon them, killed the most part, and, with much ado, a few remaining got to " their vessels, while their beloved *King* was carried, like a sack-load, upon a beast down to the river, where " there were scarce sailors enough saved from the slaughter to man the vessels."

DERRING relates, that a friend of his, a Dr. Medley, has several times eaten three or four of the berries, without receiving any hurt: and HALLER mentions his having seen a medical student swallow several. It is probable that these berries will not kill, unless many are eaten, but perhaps this poison, like many others, may act differently on different constitutions.

Vinegar has been recommended as an antidote to its poison; but powerful evacuations, particularly vomiting, are most to be depended on. In cases where a poison of this kind is known to have been swallowed, the medical practitioner will be justified in a bold practice, for his patient is not only in a very dangerous situation, but the effect of emeticks has been known to be lessened by the poison, so that fourteen grains of Emetick Tartar have been scarcely sufficient to excite vomiting.

Many substances, which in large quantities, or injudiciously administered, have proved poisonous, in small doses, skilfully exhibited, have been found extremely efficacious in the cure of diseases, and hence this, as well as other plants have been tried, particularly in such disorders as have no impression made on them by common remedies; but after numerous trials, there appears but little hopes of success from the *Atropa Belladonna.*

Such as wish to know the particular diseases against which the Deadly and the Garden Nightshades have been directed, with the various symptoms they have produced on being taken, may consult GATAKER's *Observations on the Internal Use of the Nightshade, with the Supplement; and* BROMFIELD's *Account of the English Nightshades, and their Effects,* 1757.

We have seen a goat eat, without injury, the leaves and stalks; and the caterpillar of the *Phalæna Antiqua, Roesel t. 39,* and *Braffica Roesel t. 29,* feed on its foliage.

LYCOPSIS ARVENSIS. FIELD, or SMALL WILD BUGLOSS.

LYCOPSIS *Lin. Gen. Pl.* PENTANDRIA MONOGYNIA. *Corolla tubo incurvato.*

Raii Syn. Gen. 13. HERBA ASPERIFOLIA.

LYCOPSIS *arvensis* foliis lanceolatis hispidis, calycibus florescentibus erectis. *Lin. Syst. Vegetab.* p. 160. Sp. Pl. p. 199. Fl. Suec. n. 167. Fl. Lappon. 77.

LYCOPSIS foliis asperrimis, undulatis, ferratis, linguaeformibus. *Hall. hist.* 605.

ECHIUM Fuchsii feu Borrago sylvestris. *J. B.* III. 581.

BUGLOSSUM sylvestre minus. *Bauh. pin.* 256. *Parkins.* 765. *Dillen. Nov. Gen. Tab.* 3.

BUGLOSSA sylvestris minor. *Ger. emac.* 799. *Raii Syn.* p. 227. *Hudson. Fl. Angl.* p. 82. *Lightfoot Fl. Scot.* p. 135.

RADIX annua, simplex, fibrosa, albida.

ROOT annual, simple, fibrous, and whitish.

CAULIS pedalis, et ultra, erectus, subangulosus, hispidus, plerumque superne tantum ramosus.

STALK a foot or more in height, upright, slightly angular, hispid, for the most part branched at top only.

FOLIA alterna, sessilia, lanceolata, obtusiuscula, papilloso-hispida, subtus pallidiora, venis, margine undulata, subrevoluta.

LEAVES alternate, sessile, lanceolate, bluntish, hispid, hairs issuing from small papillae, palest on the under side, veinlets, waved at the edge, and slightly rolled back.

FLORES caerulei, spicati, secundi, sessiles, deorsum spectantes.

FLOWERS blue, growing in spikes, all one way, sessile, and turned backward.

BRACTEAE foliis subsimiles.

FLORAL-LEAVES somewhat like the leaves themselves.

CALYX: PERIANTHIUM, quinquepartitum, hispidum, persistens, laciniis oblongis, acutis, longitudine fere corollae.

CALYX: a PERIANTHIUM deeply divided into five segments, hispid, and permanent; the segments oblong, pointed, and almost the length of the corolla.

COROLLA monopetala, infundibuliformis; tubus cylindraceus, curvato-flexus, fig. 2. limbo semiquinquefidus, obtusus; fauce clausa squamulis quinque, pilosis, albis, fig. 3.

COROLLA monopetalous, funnel-shaped; tube cylindrical, crooked, fig. 2. limb slightly divided into five segments, obtuse; mouth closed by five, small, white, hairy scales, fig 3.

STAMINA: FILAMENTA quinque, minima, ad flexuram tubi corollae; ANTHERAE parvae, fuscae, fig. 4.

STAMINA: five FILAMENTS, very minute, at the curvature of the tube of the corolla; ANTHERAE small and brown, fig. 4.

PISTILLUM: GERMINA quatuor, viridia, glabra; STYLUS filiformis, longitudine staminum; STIGMA obtusum, subbifidum, fig. 5.

PISTILLUM: GERMINA four, green and smooth; STYLE filiform, the length of the stamina; STIGMA obtuse and slightly bifid, fig. 5.

PERICARPIUM nullum, Calyx seu semina fovens, maximus, laciniis conniventibus donec semina nigrescant deinde patescibus.

SEED-VESSEL none, the Calyx which contains the seed in its bosom, is very large, closing together till the seeds grow black, and then spreading.

SEMINA quatuor, majuscula, nigricantia, reticulato-rugosa, acutiuscula; fig. 6.

SEEDS four, largish, nearly black, with a reticulated or wrinkly surface, and a little pointed, fig. 6.

RECEPTACULUM punctis quatuor fuscis excavatis notatum.

RECEPTACLE marked with four round dots, hollowed out.

The *Lycopsis Arvensis* is a very common plant in the corn fields, especially such as are sandy, and on dry banks, in the neighbourhood of London. We have sometimes seen it so plentiful as to be highly injurious to the husbandman; it may be found in flower from May to July.

The following account of the medicinal virtues of this plant appeared lately in most of our newspapers: without vouching for the truth of the report, we have thought it our duty to lay it before our readers, with a sincere wish that the herb may prove as efficacious in its application, as it here represented.

"The celebrated M. JEAN FONTANA, Member of the learned academy of Turin, has lately published, for the general good of suffering mankind, a specific remedy against the ANTHRAX, or corrosive ulcer, otherwise called Carbuncle, or Plague-Sore. The curative prescription was communicated to him by the person who has administered it for many years to patients of that description, and with constant success. It consists simply in the use of a field plant, called by Linnaeus, LYCOPSIS ARVENSIS. Bruise and pound the plant; lay it on the manner, fix it there by means of a bandage, and do not touch it before it hath remained twenty-four hours. During the first six or seven hours, the patient will feel a painful and burning heat in the part. It often happens that on taking off the first applied, the sough gets loose and discovers a wound, which heals in a few days, by applying to it a plaster of the unguent called Basilicon. If the case should be otherwise, the first method of cure must be repeated. This second application of the bruised plant, which will not occasion above two hours pain to the patient, will be fully sufficient to remove the slough, and then the use of the above plaster effects a speedy and radical cure."

Lycopus europaeus

Lysimachia nemorum.

LYSIMACHIA NEMORUM. WOOD MONEYWORT, or LOOSESTRIFE.

LYSIMACHIA *Linnæi Gen. Pl.* Pentandria Monogynia.

Cor. rotata. Capf. globofa, mucronata, 10-valvis.

Raii Syn. Gen. 18. Herbæ fructu sicco singulari flore monopetalo.

LYSIMACHIA nemorum foliis ovatis acutis, floribus folitariis, caule procumbente. *Lin. Syft. Vegetab.* p. 165. Sp. Pl. p. 211.

LYSIMACHIA caule decumbente, foliis ovato-lanceolatis, petiolis alaribus unifloris. *Haller hift.* p. 278.

ANAGALLIS lutea nemorum. *Bauhin Pin.* p. 252.

ANAGALLIS lutea. *Gerard emac.* 618.

ANAGALLIS flore luteo. *Parkinf.* 558.

ANAGALLIS lutea nummularìæ fìmilis. J. Bauh. III. 370. Raii Syn. p. 282. Yellow Pimpernel of the Woods. *Hudfon Fl. Ang.* p. 85. *Lightfoot Fl. Scot.* p. 138.

RADIX perennis, fibrofa, fibris albidis.

CAULES plures, decumbentes, teretiufculi, utrinque fulcati, idque alterne, læves, rubentes, ex imâ parte radicantes.

FOLIA oppofita, petiolata, ovata, acuta, utrinque glabra, fubundulata, e flavo-viridia, venis prominulis; petiolis brevibus, latiufculis.

PEDUNCULI axillares, bini five folitarii, teretes, uniflori, tenues, quam folia longiores.

CALYX: Perianthium quinquepartitum, perfiftens, laciniis fubulatis, fubtriangularibus, *fig.* 1.

COROLLA monopetala, flava, tubus nullus; limbus quinquepartitus, laciniis ovatis, *fig.* 2. 3. bafi faturatius flavis, nitidique, in fauce corollæ glandulæ flavæ inter filamenta locantur, et margo corollæ glandulis pedicellato ornatur, *fig.* 6.

STAMINA: Filamenta quinque, lævia erecta, medio paulo craffiora; Antheræ oblongæ, incurvatæ, *fig.* 4. 5.

PISTILLUM: Germen fubrotundum, læve; Stylus filiformis, apice paulo craffior; Stigma fimplex, *fig.* 7.

PERICARPIUM: Capfula globofa, unilocularis, *fig.* 8.

SEMINA plurima, orbiculata, plana, *fig.* 9.

ROOT perennial, fibrous, the fibres whitifh.

STALKS feveral, decumbent, roundifh, with a furrow on each fide, and that alternately, fmooth, of a red colour, ftriking root at the bafe.

LEAVES oppofite, ftanding on foot-ftalks, ovate, pointed, gloffy on each fide, fomewhat waved, of a yellowifh-green colour, the veins a little prominent; leaf-ftalks fhort and broadifh.

FLOWER-STALKS axillary, growing fometimes in pairs, fometimes fingly, round, one-flower'd, flender, and longer than the leaves.

CALYX a Perianthium deeply divided into five fegments, and permanent, the fegments awl-fhaped, and fomewhat triangular, *fig.* 1.

COROLLA monopetalous, **yellow**, *tube* wanting, the limb divided into five ovate fegments, *fig.* 2. 3. at bottom more intenfely yellow and fhining, in the mouth of the corolla fmall yellow glands are obfervable betwixt the filaments, and the edge of the corolla is ornamented with little glands ftanding on foot-ftalks, *fig.* 6.

STAMINA: five Filaments, fmooth, upright, fomewhat thickell in the middle; Antheræ oblong, bent a little downwards, *fig.* 4. 5.

PISTILLUM; Germen roundifh, fmooth; Style filiform, fomewhat thickell at top; Stigma fimple, *fig.* 7.

SEED-VESSEL: a globular Capfule of one cavity, *fig.* 8.

SEEDS numerous, round, and flat, *fig.* 9.

When the bloffoms of this plant are expanded, they fomewhat refemble thofe of the common Pimpernel in fhape, and hence the older Botanifts, who paid little regard to fuch minute but neceffary diftinctions, as the harinefs of the Filaments, &c. confidered it as an *Anagallis*; Linnæus has joined it with the Moneywort, to which, in its general habit, it bears no fmall affinity, but from which it effentially differs in many particulars; the leaves, for inftance, are more pointed, the flowers are fmaller, lefs bell-fhaped, and ftand on much longer foot-ftalks, and the ftalks are generally redder.

This fpecies grows in moift woods, and is not uncommon in the neighbourhood of London; in Charlton-Wood it particularly abounds, flowering from June to September.

Lysimachia vulgaris

LYSIMACHIA VULGARIS.   YELLOW LOOSE-STRIFE.

LYSIMACHIA *Lin. Gen. Pl.* PENTANDRIA MONOGYNIA.

Cor. rotata. *Capf.* globofa, mucronata, decemvalvis.

Raii Syn. Gen. 18. HERBÆ FRUCTO SICCO SINGULARI FLORE MONOPETALO.

LYSIMACHIA *vulgaris* paniculata, racemis terminalibus. *Lin. Syft. Vegetab.* p. 165. *Sp. Pl.* p. 209. *Fl. Suecic.* n. 175.

LYSIMACHIA foliis ovato-lanceolatis, fpicis paniculatis. *Hall. Hift.* 630.

LYSIMACHIA *vulgaris. Scopoli Fl. Carn.* n. 214.

LYSIMACHIA lutea. *I. B. II.* 901. *Ger. emac.* 474.

LYSIMACHIA lutea major quam Diofcoridis. *Bauh. Pin.* 245.

LYSIMACHIA lutea major vulgaris. *Park.* 544. Yellow Willow-herb or Loofe ftrife. *Raii Syn.* 282. *Hudfon Fl. Angl. ed.* 2. p. 86. *Lightfoot Fl. Scot.* p. 138.

RADIX perennis, repens.

CAULIS tripedalis et ultra, erectus, ubi folia bina obtufe tetragona, ubi terna fulcatus, feu angulofus, angulis obtufis; fuperne bifurculus, inferne glaber, ramofus, ad geniculis paululum incrafiatus.

FOLIA bina, feu terna, quaterna et quina etiam obfervavi, feffilia, ovato-lanceolata, integra, marginae inaequali, venofa, nuda.

FLORES paniculati, lutei, racemis terminalibus ex alis foliorum.

PEDUNCULI unifiori, fubvifcidi, apice incrafiati.

CALYX: PERIANTHIUM monophyllum, quinquepartitum, sexcum, crectum, perfiftens, laciniis ftriatis, rubro marginatis, specibus ante et poft florefcentiam tortuofis. *fig.* 1.

COROLLA monopetala, rotata. *Limbus* quinquepartitus, laciniis ovatis, acutis. *fig.* 2.

STAMINA: FILAMENTA quinque, inaequalia, corolla breviora, fubulata, compreffa, vifcofa, bafi coeuntia. ANTHERÆ incumbentes, fubfagittatae. *fig.* 3.

PISTILLUM: GERMEN fubrotundum. STYLUS filiformis, longitudine ftaminum, perfiftit florefcentiâ elongatus. STIGMA obtufum. *fig.* 4.

PERICARPIUM: CAPSULA globofa, mucronata, decemvalvis.

SEMINA plurima, minima.

RECEPTACULUM globofum, maximum.

ROOT perennial and creeping.

STALK three feet or more in height, when the leaves grow in pairs, obtufely four-cornered, when three together, grooved or angular, angles obtufe, the upper part of the ftalk flightly hairy, the lower fmooth, branched, and a little thickened at the joints.

LEAVES growing in pairs, or three together, I have even noticed them growing four or five together, feffile, ovate and pointed, entire but not perfectly even on the edges, veiny and deftitute of hairs.

FLOWERS yellow, forming a panicle, flower-branches terminal, growing from the alæ of the leaves.

FLOWER-STALKS fingle-flowered, fomewhat vifcid, and thickened at the extremity.

CALYX: a PERIANTHIUM of one leaf, deeply divided into five fegments, pointed, upright, and permanent, the fegments ftriated, and edged with red, the tips both before and after flowering twifted. *fig.* 1.

COROLLA monopetalous, wheel-fhaped. *Limb* deeply divided into five fegments, which are ovate and pointed. *fig.* 2.

STAMINA: five FILAMENTS, unequal, fhorter than the corolla, tapering, flattened, vifcid, growing together at bottom. ANTHERÆ incumbent, fomewhat arrow-fhaped. *fig.* 3.

PISTILLUM: GERMEN roundifh. STYLE filiform, the length of the ftamina, lengthened out as the flowers go off. STIGMA blunt. *fig.* 4.

SEED-VESSEL a globular capfule of one cavity, and ten valves.

SEEDS numerous, very minute.

RECEPTACLE globular, and very large.

Some of the ancient writers attributed a very fingular property to this plant; no lefs than a power of taming ferocious, and reconciling difcordant animals; and hence they derive its name of *Lyfimachia* *. Others attribute the origin of its name to the learned and brave LYSIMACHUS, who, they fay, was its firft difcoverer: however this be, our Englifh name of *Loofe-ftrife* appears evidently to be founded on the power thus alfo afcribed to it.

This herb, though not fo common as its name feems to imply, is tolerably frequent about *London*, in moift meadows, and by water-fides, efpecially in the environs of the *Thames*.

It varies much in the number of the leaves at the joints, and confequently in the angular appearance of its ftalk. The twifted tips of the Calyx, though very remarkable, do not appear to have been noticed by authors.

Such as wifh to ornament the edge of a river, or piece of water, cannot felect a more proper place; but its beautiful effect will be heightened by planting with it the *Lythrum Salicaria*, both of thefe have ftrong perennial roots, and will alfo readily grow in gardens where the foil is moift.

It flowers in *July* and *Auguft*.

Some afcribe to it the power of dying green.

---

* A potfon deftructive for bne the edge of common farmers, of taking away ftrife or debate between beafts, ... only that it fo conce together, but even thofe that are wild alfo, by causing them tame and quiet, which, as they fay, this herb will do, if it be done put them each yoke or their necks, which have true I leave to them who fhall try and find it fo. *Parkin.* p. 544.

CHENOPODIUM OLIDUM. STINKING BLITE, or ORACH.

CHENOPODIUM *Lin. Gen. IV.* PENTANDRIA DIGYNIA.

Cal. 5-phyllus, 5-gonus. *Cor.* o. *Semen* 1. lenticulare superum.

Raii Syn. Gen. 5. HERBÆ FLORE IMPERFECTO SEU STAMINEO VEL APETALO POTIUS.

CHENOPODIUM *Vulvaria* foliis integerrimis, rhomboideo-ovatis, floribus conglomeratis axillaribus. *Lin. Syst. Vegetab. p.* 216. *Sp. Pl.* 321. *Fl. Suec.* 222.

CHENOPODIUM caule diffuso, foliis obtuse lanceolatis. *Haller hist. n.* 1577.

CHENOPODIUM *Vulvaria. Scopoli Fl. Carn. n.* 281.

ATRIPLEX foetida. *Bauh. Pin.* 119.

ATRIPLEX olida. *Ger. emac.* 327.

ATRIPLEX sylvestris foetida. *Park.* 749.

BLITUM foetidum Vulvaria dictum. *Raii Syn. p.* 156. Stinking Orache. *Hudson Fl. Angl. ed.* 2. *p.* 107. *Lightfoot Fl. Scot. p.* 149.

Tota planta farina alba pellucida adspersa.	The whole plant sprinkled with a white pellucid meal.
RADIX annua, fibrosa.	ROOT annual and fibrous.
CAULES plures, diffusi, teretes, subfibrati, nudiusculi.	STALKS numerous, spreading, round, somewhat striated, and thinly beset with leaves.
FOLIA alterna, petiolata, rhomboideo-ovata, integerrima.	LEAVES alternate, standing on footstalks, rhomboid-ovate, perfectly entire.
FLORES axillares et terminales, dense glomerati, subspicati.	FLOWERS axillary and terminal, thickly clustered, and somewhat spiked.
FRUCTIFICATIO a reliquis hujus generis vix diversa.	FRUCTIFICATION scarcely different from the rest of this genus.
Fig. 1. exhibet Calycem, Stamina, cum Pistillo.	*Fig.* 1. exhibits the Calyx, with the Stamina and Pistillum.
Fig. 2. Semen Calyce inclusum.	*Fig.* 2. The Seed enclosed by the Calyx.
Fig. 3. Semen seorsim. Omnia auct.	*Fig.* 3. The Seed separate. All magnified.

There is some difficulty in ascertaining several of the plants of this genus, but that difficulty cannot be alleged against the present species, as it is at all times, both fresh and dried, discoverable by its smell alone: the whole plant, if ever so lightly bruised betwixt the thumb and fingers, communicating a very permanently disagreeable odour, resembling, in the opinion of most persons, stale salt fish: it is, moreover, a procumbent plant.

This species is very common in the neighbourhood of London, **on dry banks, and at the foot of walls and** palings, where it flowers from July to September. Lewis errs egregiously when he says it naturally delights in moist places.

It is a plant of little consequence, except in a medicinal point of view, and in that its virtues are, perhaps, ill-founded; it retains, however, a place in the London and Edinburgh Dispensatories.

" Stinking Orache, on account of its strong scent, is reckoned an useful antihysteric; in which intention,
" some recommend a conserve of the leaves, others a watery infusion, and others a spirituous tincture of
" them. On some occasions it may, perhaps, be preferable to the foetids, which have been more commonly
" made use of, as not being accompanied with any pungency or irritation, and seeming to act merely by
" virtue of its odorous principle." *Lewis's Mat. Med. p.* 124.

Chenopodium olidum.

SCANDIX PECTEN. SHEPHERDS NEEDLE, or VENUS'S COMB.

SCANDIX *Lin. Gen. Pl.* PENTANDRIA DIGYNIA.

Corolla radiata. Fructus fubulatus. Petala emarginata. Flofculi difci faepe mafculi.

Raii Syn. Gen. 11. UMBELLIFERAE HERBAE.

SCANDIX *Peften feminibus laevibus roftro longiffimo. Lin. Syft. Veget. ed.* 14. *p.* 287. *Sp. Pl. p.* 368.

MYRRHIS feminis corona longiffimo. *Haller hift. n.* 751.

SCANDIX *Peften. Scopoli Fl. Carn. n.* 319.

SCANDIX femine roftrato vulgaris. *Bauh. Pin.* 152.

PECTEN VENERIS I. B. III. 2. 71.

PECTEN VENERIS feu fcandix. *Ger. emac. p.* 1040.

SCANDIX vulgaris, feu Peften Veneris. *Park.* 916. *Raii Syn. p.* 207. Shepherds Needle, or Venus's Comb. *Hudfon Fl. Angl. ed.* 2. *p.* 123. *Lightfoot Fl. Scot. p.* 162. *Jacquin Fl. Auftr. t.* 263.

RADIX annua, fimplex, albida, paucis fibrillis inftrufta. | ROOT annual, fimple, whitifh, furnifhed with few fibres.

CAULIS nunc folitarius, nunc plures ex eadem radice, ramofi, diffufi, villofi, femipedales, aut pedales, inferne purpurei, aut lineis purpureis ftriati, teretes, ad geniculos vix incraffati. | STALK fometimes fingle, fometimes feveral from the fame root, branched, fpreading, villous, half a foot or a foot in height, below purple, or ftriped with purple lines, round, and fcarcely thickened at the joints.

FOLIA dauci inftar tenuiter divifa, ad bafin vaginantia, laciniis linearibus, bifidis trifidifve, acutis, ad lentem rariter ciliatis, fig. 1. | LEAVES finely divided like thofe of wild carrot, forming a fheath at bottom, fegments linear, bifid or trifid, pointed, and, if viewed with a microfcope, thinly edged with hairs, fig. 1.

INVOLUCRUM univerfale nullum. | INVOLUCRUM: general Involucrum wanting.

UMBELLA: univerfalis plerumque biradiata. | UMBEL: general Umbel ufually compofed of two radii.

INVOLUCRUM partiale magnum, pentaphyllum, foliolis nervofis, ciliatis, bifidis. | INVOLUCRUM: partial Involucrum large, five-leaved, leaflets rib'd, edged with hairs, and bifid.

FLORES Umbellulae quinque ad feptem, plerumque fertiles, albae. | FLOWERS of the fmall Umbel from five to feven, for the moft part fertile and white.

COROLLA: Petala quinque, obverfe ovata, apice inflexa, patentia, exteriore majore, fig. 2. | COROLLA: five Petals, inverfely ovate, bent in at the tip, fpreading, the outermoft petal largeft, fig. 2.

STAMINA: Filamenta quinque, alba; Antherae primo virefcentes, demum nigricantes, fig. 3. | STAMINA five white Filaments; Antherae firft greenifh, finally blackifh, fig. 3.

PISTILLUM: Germen breviffime pedicellatum, oblongum, hirfutulum; Styli duo, fubulati, erefti, perfiftentes; Stigmata fimplicia, fig. 4. 5. | PISTILLUM: Germen ftanding on a very fhort footftalk, oblong and flightly hairfute, Styles two, tapering, upright and permanent; Stigmata fimple, fig. 4. 5.

SEMINA duo, fufca, hinc convexa, ftriata, inde plana hirfotula, in roftrum longiffimum excurrentia, fig. 7. | SEEDS two, brown, convex and ftriated on one fide, and flat on the other, flightly hairfute, running out into a very long beak, fig. 7.

NECTARIUM: ad bafin ftylorum, purpurei coloris, fig. 6. | NECTARY at the bafe of the ftyles, of a purple colour, fig. 6.

Common in corn fields, not only in Great-Britain, but in all the fouthern parts of Europe, fometimes fo plentiful, as to prove injurious to the farmer.

Is particularly diftinguifhed from all our other umbelliferous plants by the uncommon length of the beak of the feeds, as well as by the fingularity of the leaves of the Involucellum, which are uncommonly large and bifid.

Flowers in June, and ripens its feed in July.

Its feed-leaves, on their firft appearance above ground, are uncommonly long.

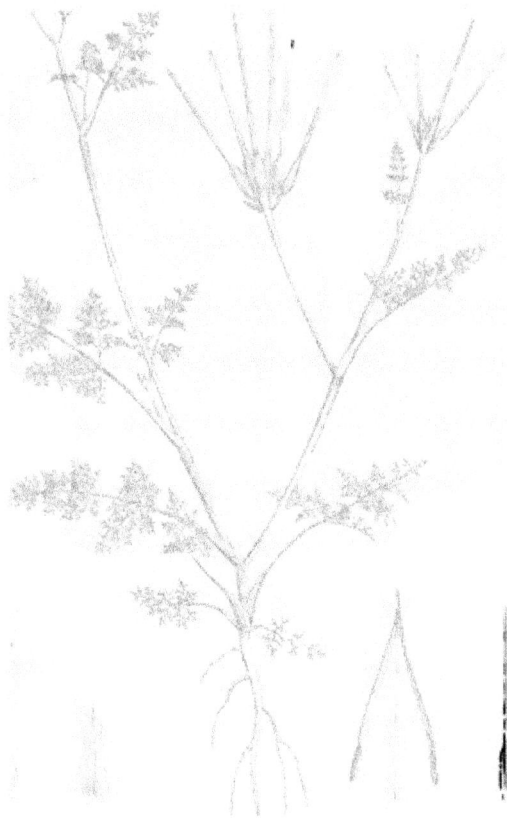

LINUM USITATISSIMUM. COMMON FLAX.

LINUM *Lin. Gen. Pl.* PENTANDRIA PENTAGYNIA.

Cal. 5-phyllos. *Petala* 5. *Capf.* 5-valvis, 10-locularis. *Sem.* folitaria.

Roii Syn. Gen. 24. HERBÆ PENTAPETALÆ VASCULIFERÆ.

LINUM *ufitatiffimum* calycibus capfulifque mucronatis, petalis crenatis, foliis lanceolatis alternis, caule fubfolitario. *Lin. Syft. Vegetab. p.* 249. *Sp. Pl. p.* 397.

LINUM *arvenf. Bauh. Pin.* 214.

LINUM *fylveftre vulgatius. Park.* 1334. *Ger. emac.* 556. *Roii Syn. p.* 362. Manured Flax. *Hudfon. Fl. Angl. ed.* 2 *p.* 133. *Lightfoot Fl. Scot. p.* 173.

RADIX annua, fimplex, fibrofa, pallide fufca.

CAULIS erectus, fefquipedalis, bipedalis et ultra, teres, glaber, foliofus, fuperne tantum ramofus.

FOLIA lanceolata, feffilia, conferta, fparfa, fuberecta, integerrima, lævia, trinervia.

FLORES majufculi, pulchre-cærulei, paniculati.

PEDUNCULI teretes, glabri.

CALYX: PERIANTHIUM 5-phyllum, foliolis ovatis, acuminatis, carinatis, perfiftentibus, margine membranaceis, ad lævem ciliatis, *fig.* 1.

COROLLA: PETALA 5, cærulefcentia, cuneifolia, decidua, venis faturatioribus picta, unguibus albis, apicibus tuberofis, *fig.* 2.

STAMINA: FILAMENTA quinque, alba, fubulata, bafi aliptata. ANTHERÆ primo oblongæ, demum fagittæ, *fig.* 3. incumbentes, cæruleæ, ad ftylos inclinatæ et fubconduntæ, *fig.* 3. 4.

PISTILLUM: GERMEN ovatum, nitidens. STYLI quinque, longitudine filamentorum, fub-clavati, cærulefcentes, apice leviter cohærentes. STIGMATA fimplicia, *fig.* 5.

PERICARPIUM: CAPSULA globofa, fubangulata, mucronata, decemlocularis, quinquevalvis, *fig.* 6.

SEMINA in fingulo loculamento folitaria, ovato-acuta, compreffa, nitida, *fig.* 7.

ROOT annual, fimple, fibrous, of a pale brown colour.

STALK upright, a foot and a half, two feet high or more, round, fmooth, leafy, branched above only.

LEAVES lanceolate, feffile, growing thickly together, without any regular order, almoft upright, perfectly entire.

FLOWERS large, of a beautiful blue colour, growing in a panicle.

FLOWER-STALKS round and fmooth.

CALYX: a PERIANTHIUM of five leaves, which are ovate, pointed, keeled, permanent, the edge membranous, and if magnified fringed with hairs, *fig.* 1.

COROLLA: 5 blueifh, wedge-fhaped, deciduous PETALS, ftreaked with veins of a deeper colour, claws white, tips fomewhat gnawed, *fig.* 2.

STAMINA: five white tapering FILAMENTS, dilated at the bafe. ANTHERÆ at firft oblong, finally arrow-fhaped, *fig.* 3. incumbent, of a blue colour, inclined to the ftyles, and fomewhat united, *fig.* 3. 4.

PISTILLUM: GERMEN ovate, fhining. STYLES five, the length of the filaments, fomewhat club-fhaped, blueifh, flightly cohering. STIGMATA fimple, *fig.* 5.

SEED-VESSEL: a globular, fomewhat angular and pointed CAPSULE, having ten cavities, and five valves, *fig.* 6.

SEEDS one in each cavity, ovate, pointed, flat and gloffy, *fig.* 7.

It may be doubted, perhaps, whether the common flax, found in any part of the kingdom, may not originally have been introduced from abroad; yet Mr. HUDSON fpeaks of it as a common plant in Dorfetfhire and Devonfhire, and entertains no idea of its being a doubtful native. However this may be, the few fpecimens of it which we find occafionally in corn fields and among rubbifh, particularly in the neighbourhood of Batterfea (for flax is not cultivated near London), have doubtlefs been introduced there with the produce of the garden or the corn field.

It flowers in June and July.

In the earlieft record we have, flax is mentioned as a plant cultivated in Egypt (Exodus ch. ix. v. 31.); for which reafon antiquaries have been furprifed to find the veftments of mummies made of cotton. It is highly probable, however, that mankind made ufe of cotton before the ufe of flax was difcovered; for cotton is produced in a ftate ready for fpinning, whereas flax requires a long procefs before it can be brought to that ftate.

In the fimplicity of former times, when families in this ifland provided within themfelves moft of the neceffaries and conveniencies of life, every garden fupplied a proper quantity of hemp and flax; but the macerating or fteeping, which was neceffary to feparate the thread by rotting the ftalk, was in many places found to render the water fo offenfive and detrimental, that in the reign of Henry VIII. a law was made that " No perfon fhall water " any hemp or flax in any river, running water, ftream, brook, or other common pond, where beafts are ufed to be watered, " on pain of forfeiting, for every time fo doing, twenty fhillings. 33 Hen. VIII. c. 17. § 1. Might not this inconvenience be prevented, and the procefs much accelerated, by ufing boiling water, and a proper quantity of the afhes of any vegetable? *Vid.* below.

The wifdom of Parliament hath lately thought proper to encourage, by a premium, the growth of hemp and flax in this kingdom, certainly with a very laudable intention, as long as we procure thefe articles from countries where the balance of trade is againft us; or, in other words, while we contrive to pay for them in money, and not with our manufactures. The premium is four pence for every fourteen pounds of flax.

The ancients were of opinion, that flax impoverifhed land. " Urit enim lini campum feges." *Virg.* G. I. v. 77. But, while fpeculative and practical cultivators unfortunately continue to be fuch very diftinct people, the rules which we find in books cannot be much depended on. However, it may be a caution to thofe who have not a plentiful command of manure not to engage too largely with this plant without proper trials. As flax will be new

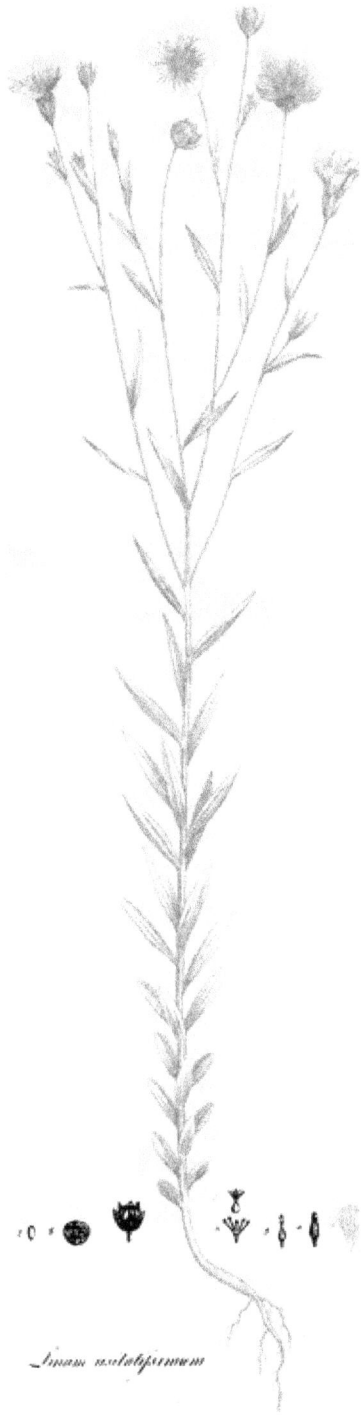

Linum usitatissimum

to most of the land in the kingdom, there is little doubt but that the produce will at first be large, and it is very desirable to introduce a new kind of grass into husbandry to extend the rotation of crops.

> " *For the vicissitude of various grain*
> " *Tend to prepare the vigour of the plain.*"

Flax not only supplies us with cloathing, but its seeds, well known by the name of lin-feed, afford an oil of great use in painting, varnishing, &c. They are also used medicinally. Infusions of lin-feed, like other mucilaginous liquors, are used as emollients, incrassants, and obtunders of acrimony, in heat of urine, stranguries, thin defluxions on the lungs, and other like disorders. A spoonful of the seeds, embruised, is sufficient for a quart of water, larger proportions rendering the liquor disagreeably slimy. The mucilage obtained by infusing the infusions or decoctions is an excellent addition for reducing disgustful powders into the form of an electuary, occasioning the compound to pass the fauces freely, without sticking or discovering its taste in the mouth. The expressed oil is supposed to be more of a healing and balsamic nature than the other oils of this class, and has been particularly recommended in coughs, spitting of blood, cholics, and constipations of the belly. The seeds in substance, or the matter remaining after the expression of the oil, are employed externally in excellent and maturating cataplasms. In some places these seeds in times of scarcity have supplied the place of grains but appeared to be so unwholesome as well as an unpalatable food. Travel relates, that those who fed on them in Zealand had the hypochondria in a short time distended, and the face and other parts swelled; and that not a few died of these complaints.

The following reflections communicated to me by a friend will, I flatter myself, not be unacceptable to my readers. Should practice justify the theory, I will venture to say, they will be golden reflections to the nation.

Some reflections relative to the watering of flax by a new method, so as to shorten labour, add to the strength of the flax, and give it a much finer colour, which would render the operation of bleaching safer and less tedious.

THOUGH the following reflections have for their object an improvement in the very essential article of watering of flax, yet I must advertise my reader, that they are only theory, and must depend entirely for their truth and justification upon future experiments, skilfully and judiciously made. Should repeated trials prove the advantage of the method proposed, we may venture to affirm, it would be an improvement that would increase the national income in the agricultural branch many thousand pounds annually, would add greatly to the perfection of the linen manufacture, and over and above would suppress a very disagreeable nuisance, which the present method of watering flax occasions during some part of the summer in every flax-growing country.

The intention of watering flax is, in my opinion, to make the boon more brittle or fissile, and by soaking to dissolve that gluey kind of sap that makes the bark of plants and trees adhere, in a small degree, to the woody part. The bark is called the harle, and produces the flax; the useless woody part, which remains when the bark is separated, the boon. To effect this separation easily, the practice has long prevailed of soaking the flax in water to a certain degree of fermentation, and afterwards drying it. For this soaking some prefer rivulets that have a small current, and others stagnant water in ponds and lakes. In both these ways the water acts as in all other cases of infusion and maceration. After two or three weeks it extracts a great many juices of a very strong quality, which in ponds give the water an inky tinge, and offensive smell, and in rivulets what is in the stream, and kill the fish.

Nay, if this maceration is too long continued, the extracted and fermented sap will completely kill the flax itself: for if, instead of two or three weeks, the new flax were to lay soaking in the water four or five months, I presume it would be good for nothing but to be thrown upon the dunghill. Both harle and boon would in that time be completely rotted; yet the harle or flax, when entirely freed from this sap, and manufactured into linen, or into ropes, might be many months under water without being much damaged. As linen, it may be washed, steeped, and boiled in scalding water twenty times, without losing much of its strength; and as paper, it acquires a kind of incorruptibility.

It appears then essential, to the right management of new flax, to get rid of this pernicious vegetative sap, and to macerate the boon; but from the complaints made against both the methods of watering now in use, there is reason to think, that there is still great room for improvement in that article. In rivulets, the vegetative sap, as it is dissolved, is carried off by the current, to the destruction of the fish. This prevents the flax from being stained; but the operation is tedious, and, I have been told, often not complete, from the uncertainty of knowing the precise times when it is just enough, and not too much, or perhaps from neglect. In ponds, the inky tinge of the water often serves as a kind of dye to the flax, which imbibes it so strongly, that double the labour in bleaching will hardly bring the linen made of such flax to an equality in whiteness with linen made of flax unfinged. This seems to be equally unwise, as though we were to dye cotton black first, as a means to whiten it afterwards. These ponds besides become a great nuisance to the neighbourhood: the impregnated water is often of such a pernicious quality, that cattle, however thirsty, will not drink of it, and the effluvia of it may perhaps be nearly as infectious as it is offensive. If this effluvia is really attended with any contagious effects in our cold climate, a thing worth enquiring into, how much more pernicious must its effects have been in the hot climate of Egypt, a country early noted for its great cultivation of flax!

From these considerations I have been led to think, that the process of watering might be greatly improved and shortened by plunging the new flax, after it is rippled, into scalding water, which, in regard to extracting the vegetative sap, would do in five minutes more than cold water would do in a fortnight, or perhaps more than cold water could do at all, in respect to the clearing the plant of that sap. Rough almonds, when thrown into scalding water, are blanched in an instant; but perhaps a fortnight macerating these almonds in cold water would not make them part so easily with their skins, which are the same to them as the harle to the flax. Were tea leaves to be infused in cold water a fortnight, perhaps the tea produced by that infusion would not be so good to the taste, nor so strongly tinged to the eye, as what is effected by scalding water in five minutes. By the same analogy, I think, flax, or any small twig, would be made to part with its bark much easier and quicker, by being dipped in boiling water, than by being steeped in cold water. This reflection opens a door for a great variety of new experiments in regard to flax. I would therefore recommend to gentlemen cultivators and farmers to make repeated trials upon this new system, which would soon ascertain whether it ought to be adopted in practice or rejected. One thing, I think,

think, we may be certain of, that, if the Egyptians watered their flax in our common manner, they undoubtedly watered it in very warm water, from the great heat of their climate, which probably might make them neglect to think of water heated by any other means than that of the sun. A good general practice can only be established upon repeated trials; but, I am persuaded, many lose half the value of their crop by some of the present methods of watering it. Though one experiment may fail, another with a little variation may succeed, and the importance of the object desired to be obtained will justify a good degree of perseverance in the prosecution of the means. In this view, as the Chinese thread is said to be very strong, it would be worth while to be acquainted with the practice of that distant nation in regard to the curing and manufacturing of flax, as well as with the methods used by the Flemings and the Dutch.

Boiling water perhaps might at once clear the new flax from many impurities, which, when not removed till spun into yarn, are then removed with difficulty, and loss of substance to the yarn. Why should not the longitudinal fibres of the flax, before they be spun into yarn, be made not only as fine but as clean as possible? Upon the new system proposed, the act of bleaching would begin immediately after the rippling of the flax; and a little more than might save much of what is generally done after the spinning and weaving. To spin dirty flax, with a view of cleaning it afterwards, appears to be the same impropriety as though we were to reserve part of the dressing given to leather till after it is made into a glove.

Should the plunging the flax into the boiling water not suffice to make the boon brittle enough, as I am inclined to think it would not, then the common watering might be added; but, in this case, probably half the time usually given to the watering would suffice, and the flax might then be laid in clear rivulets, without any apprehensions of its infecting the water and poisoning the fish, as of being discoloured itself; for the boiling water into which it had been previously put would have extracted all the poisonous vegetable sap, which, I presume, is what chiefly discolours the flax, or kills the fish.

On the supposition that boiling water, in the preparation of flax, may be found to be advantageous and profitable, I can recollect at present but one objection against it being generally adopted. Every flax-grower, it may be said, could not be expected to have conveniencies for boiling water sufficient for the purpose; the consumption of water would be great, and some additional expence would be incurred. In answer to this I shall only observe, that I presume any additional expence would be more than reimbursed by the better marketable price of the flax; for otherwise any new improvement, if it will not quit cost, must be dropped, were it even the finishing after gold. In a large cauldron a great deal of flax might be dipped in the same water, and the consumption perhaps would not be more than a quart to each sheaf: even a large boiler pot would be capable of containing one sheaf after another; and I believe the whole objection would be obviated, were the practice to prevail with us, as in Flanders and Holland, that the flax-grower and the flax-dresser should be two distinct professions.

I shall conclude with recommending to those who are inclined to make experiments, not to be discouraged by the failure of one or two trials. Perhaps the flax, instead of being just plunged into the scalding water, ought to be kept in it five minutes: perhaps a quarter of an hour; perhaps a whole hour. Should five minutes, or a quarter of an hour, or an hour, not be sufficient to make the boon and harle easily separate, it might perhaps be found expedient to boil the flax for more than an hour: and such boiling, when in this state, might in return save several hours boiling in the article of bleaching. It is not, I think, at all probable, that the boiling of the flax with the boon in it would prejudice the harle; for, in the course of its future existence, it is made to be exposed twenty or forty times to this boiling trial, and, if not detrimental in the one case, it is to be presumed it would not be detrimental in the other. Perhaps after the boiling it would be proper to pile up the flax in one heap for a whole day, or for half a day, to occasion some fermentation, or perhaps, immediately after the boiling, it might be proper to wash it in cold water. The great object, when the flax is pulled, is to get the harle from the boon with as little loss and damage as possible; and if this is accomplished in a more complete manner than usual, considerable labour and expence will be saved in the future manufacturing of the flax. On this account, I think, much more would be gained than lost, were the two or three last inches of the roots of the flax to be chopped off, or clipped off, previous to its being either watered or boiled.

The following precaution is necessary to be observed, that the flax should never be spread out to dry at a season when it may be in danger of being exposed to the frost.

LEUCOJUM ÆSTIVUM.   SUMMER SNOWFLAKE.

LEUCOJUM *Lin. Gen. Pl.* HEXANDRIA MONOGYNIA.

Cor. campaniformis, 6-partita, apicibus incrassata. *Stigma* simplex.

Raii Syn. Gen. 26. HERBÆ RADICE BULBOSA PRÆDITÆ.

LEUCOJUM *æstivum* spatha multiflora, stylo clavato. *Lin. Syst. Vegetab. p.* 316. *Sp. Pl. p.* 414. *Jacquin Fl. Austr. t.* 203. *v.* 4.

LEUCOJUM *æstivum. Scopoli Fl. Carn. n.* 393.

LEUCOJUM bulbosum majus f multiflorum. *Bauh. Pin.* 55.

LEUCOJUM bulbosum serotinum majus 1. *Cluf. hist.* 1. *p.* 170.

LEUCOION bulbosum polyanthemum. *Dodon. Stirp hist. p.* 230. The great late flowering Bulbous Violet. *Park. Parad. p.* 110.

RADIX: *Bulbus* magnitudine nucis castaneæ, subovatus, extus pallide fuscus, intus albus, tunicatus, lamellis plurimis, tenuibus, dense compactis.

ROOT: a Bulb the size of a chesnut, somewhat ovate, externally of a pale brown colour, internally white, coated, the coats numerous, thin, and closely compacted.

FOLIA plurima, sesquipedalia, erecta, sublinearia, saturate viridia, unciam fere lata, obtusa, superne plana, inferne leviter carinata, carina obtusa, exteriora breviora.

LEAVES numerous, about a foot and a half in length, upright, nearly linear, of a deep green colour, almost an inch in breadth, obtuse, above flat, beneath slightly keeled, the keel obtuse, the lowermost leaves shortest.

SCAPUS foliis paulo altior, multiflorus, fistulosus, subcompressus, anceps, subtortuosus, uno latere nonnunquam obtuso, altero acuto.

STALK a little higher than the leaves, supporting many flowers, hollow, slightly flattened, two-edged, a little twisted, one side sometimes obtuse, the other acute.

PEDUNCULI plerumque quinque ex eadem spatha, uniflori, angulati, longitudine inæquales.

FLOWER-STALKS for the most part five proceeding from the same sheath, each supporting a single flower, angular, and of unequal lengths.

FLORES albi, penduli, secundi, vix odori.

FLOWERS white, pendulous, growing all one way, with little scent.

COROLLA campaniformi-patens, Petala sex, ovata, alba, intus striata, basi eximie cohærentia, apicibus crassiusculis, strictioribus, macula viridi insignita.

COROLLA somewhat bell-shaped, spreading, Petals six, ovate, white, finely grooved within side, not at all uniting at bottom, tips thickish, a little puckered, and marked with a green spot.

STAMINA: FILAMENTA sex, alba, filiformia: ANTHERÆ oblongæ, subquadrangulares, erectæ, luteæ, apice poris duobus dehiscentes, *fig.* 1, 2.

STAMINA six white, thread-shaped FILAMENTS: ANTHERÆ oblong, somewhat quadrangular, upright, yellow, each cell open at top, *fig.* 2.

PISTILLUM: GERMEN subovatum, infernum Stylus albus, staminibus paulo longior, inferne attenuatus, superne virescens; STIGMA breve, setaceum, erectum, acutum, *fig.* 3.

PISTILLUM GERMEN somewhat ovate, beneath: STYLE white, a little longer than the stamina, tapering downwards, above greenish; STIGMA like a small, short, upright, pointed bristle, *fig.* 3.

PERICARPIUM: CAPSULA subpyriformis, membranacea, trilocularis, trivalvis, *fig.* 4.

SEED-VESSEL: a CAPSULE somewhat pear-shaped, membranous, having three cavities and three valves, *fig.* 4.

SEMINA plura, majuscula, subrotunda, atra, nitentia, *fig.* 5.

SEEDS several, somewhat large, nearly round, black, and glossy, *fig.* 5.

Flowers about the middle of May.

It is found undoubtedly wild, betwixt *Greenwich* and *Woolwich*, about half a mile below the former, close by the *Thames* side, just above high water mark, growing (where no garden, in all probability, could ever have existed) with *Arundo Phragmites, Caltha palustris, Oenanthe crocata,* and *Angelica sylvestris* - Prof. JACQUIN, who figures it in the *Flora Austriaca,* and SCOPOLI, in his *Flora Carniolica,* describe it as growing in similar situations; their words are, *crescit in pratis udis et sub palustribus.* It has also been found in the *Isle of Dogs,* which is the opposite shore.

How so ornamental a plant, growing in so public a place, could have escaped the prying eyes of the many Botanists who have resided in London for such a length of time, seems strange: for my own part, I am perfectly satisfied of its being a native of our island, and have no doubt but it will be found in many other parts of it.

The figure we have given, was drawn on the spot above described, where it grows more luxuriantly than we usually see it in gardens; the reason of which is, that in gardens it seldom has a soil or situation sufficiently moist.

The older Botanists, and even TOURNEFORT, united it with the Snowdrop; and in our gardens it is generally known by the name of the great Summer Snowdrop; but as it differs very essentially in its fructification from the *Galanthus,* we have thought it necessary to give it the new English name of Snowflake, to correspond in some degree with the Linnæan generic name *Leucojum.*

Leucojum aestivum.

Convallaria majalis.

CONVALLARIA MAJALIS. LILY OF THE VALLEY.

CONVALLARIA *Lin. Gen. Pl.* HEXANDRIA MONOGYNIA.

Cor. sexfida. Bacca maculosa 3-locularis.

Raii Syn. Gen. 16. HERBÆ BACCIFERÆ.

CONVALLARIA *majalis scapo nudo. Lin. Syst. Vegetab.* p. 275. *Spec. Plant.* p. 451. *Flor. Suec.* n. 292.

POLYGONATUM *scapo diphyllo, floribus bifaciatis, nutantibus, campanulatoriobus. Haller. Hist.* n. 1241.

CONVALLARIA *majalis. Scopoli Fl. Carn.* n. 418.

LILIUM convallium. *Bauh. Pin.* p. 304.

LILIUM convallium. *Ger. Emac.* p. 410. floc. albo. *Parkins. Parad.* p. 519. *Raii Syn.* p. 264. Lily-convally or May Lily. *Hudson. Fl. Angl. ed.* 3. p. 146. *Lightfoot. Fl. Scot.* p. 182.

RADIX perennis, fibrosa, fibris plurimis, teretibus, transversim rugosis, horizontaliter paulo infra terram in longum excurrit, repentibus.	ROOT perennial, fibrous, fibres numerous, round, transversely wrinkled, extending horizontally just below the surface of the earth, and creeping to a considerable distance.
SQUAMÆ quatuor, vel quinque, fupervariæ, purpurafcentes, alternæ, bafin foliorum et fcapi obvelfitant et colligant.	SCALES four or five flightly ridged, purplifh, alternate fcales furround and bind together the bafe of the leaves and ftalk.
FOLIA bina, petiolata, ovato, utrinque acuta, erecta, lævia, nervofa, altero plerumque majori, latiufcula, petiolis teretibus, exteriore paulo rubris adjecto, tubuloso ad involvendum interiorem folidum.	LEAVES growing two together, ftanding on footftalks, pointed at each end, upright, fmooth ribbed, one generally larger than the other, of a bright green colour, rootftalks round, the outermoft dotted with red, and tubular to receive the inner one which is folid.
SCAPUS lateralis, longitudine foliorum, erectus, nudus, lævis, femicylindraceus.	STALK lateral, the length of the leaves, upright, naked, fmooth, femicylindrical.
BRACTÆA lanceolata, membranacea, fub fingulo pedunculo, pedunculo brevior.	FLORAL-LEAF lanceolate, membranous, under each flower-ftalk, fhorter than the flower-ftalk.
FLORES fex, five octo, excefoch, nutantes, albi feu lutefcentes, odorati.	FLOWERS fix or eight, growing in a racemus, hanging down, white or yellowifh, and fweet-fcented.
PEDUNCULI uniflori, teretes, filiformes.	FLOWER-STALKS one flowered, round, and filiform.
CALYX nullus.	CALYX wanting.
COROLLA monopetala, globofo-campanulata. *Limbus* fexfidus, laciniis obtufiufculis, reflexis, *fig.* 1.	COROLLA monopetalous, roundifh, bell-fhaped. The *Limb* divided into fix obtufe reflexed fegments, *fig.* 1.
STAMINA: FILAMENTA fex, fubulata, petalo inferta, corolla breviora. ANTHERÆ oblongæ, erectæ, biloculares, flavæ, longitudine filamentorum, *fig.* 2.	STAMINA: fix FILAMENTS tapering, inferted into the petal, and fhorter than the corolla. ANTHERÆ oblong, upright, bilocular, yellow, the length of the filaments, *fig.* 2.
PISTILLUM: GERMEN fubrotundum, viride. STYLUS filiformis, ftaminibus longior. STIGMA obtufum, trigonum, *fig.* 3.	PISTILLUM: GERMEN roundifh, green. STYLE filiform, longer than the ftamina. STIGMA obtufe, and three-cornered, *fig.* 3.
PERICARPIUM: BACCA globofa, majufcula, rubra, trilocularis, polyfperma, *fig.* 4.	SEED-VESSEL a round, largifh, red BERRY, having three cavities, and containing many feeds, *fig.* 4.
SEMINA quinque et ultra majufcula, lutefcentia, hinc convexa, inde plana feu angulata, *fig.* 5. 6.	SEEDS five and more, largifh, yellowifh, convex on one fide, and flat or angular on the other, *fig.* 5. 6.

LINNÆUS, in his *Flora Lapponica*, p. 80. gives his reafons at large for uniting in one genus the *Lilium convallium*, the *Polygonatum*, and *Unyedium*, and for adopting the name *Convallaria*.

The Lily of the Valley claims our notice as an ornamental and a medicinal plant. As an ornamental one, few are held in greater eftimation; indeed, few are the flowers which can boaft fuch delicacy with fuch fragrance; fortunately it is moft eafy of cultivation, acquiring only to be placed on the fhady part of a garden, and to be tranfplanted now and then, when the roots are too much matted together to produce flowers freely. It bears forcing tolerably in pots, and hence the curious may have it in bloffom at leaft two months in the year.

There is a variety of it with reddifh flowers and double bloffoms. In its wild ftate it is feldom feen in berry; but gardeners thefe readily when cultivated. Like many of thofe plants which are eagerly fought after, it is now become rather fcarce in the neighbourhood of London. In Mr. RAY's time it grew plentifully on Hampftead-Heath, but is now fparingly found there. In Lord Mansfield's wood, near the Spaniard, it may be met with in greater abundance; nor is it uncommon in the woods about Dulwich. It flowers in May and June.

The flowers readily impart their fragrance, both to watery and fpirituous menftrua. Their odorous matter, like that of the white Lily, is very volatile, being totally diffipated in exficcation, and elevated in diftillation; nor does the diftilled fpirit tura milky on the admixture of water, as thofe fpirits do which are impregnated with actual oil. The pungency and bitternefs, on the other hand, refide in a fixed matter, which remains entire both in the watery and fpirituous extracts, and which in this concentrated ftate approaches, as CARTHEUSER obferves, to hepatic Aloes.

It is principally from the volatile parts of thefe flowers, that medicinal virtues have been expected in nervous and and exanthmous diforders; but probably their fix parts alfo, which have no fmell, have perhaps the greateft fhare in their efficacy. The flowers, dried and powdered, and thus divefted of their odoriferous principle, prove ftrongly fternutatory. Watery or fpirituous extracts made from them, given in dofes of a fcruple or half a dram, act as gentle ftimulating aperients and laxatives, and feem to partake of the purgative virtue as well as of the bitternefs of Aloes.

The roots poffefs a greater degree of bitternefs, and a fimilar purgative quality. *Lewis's Mat. Med.*

JUNCUS PILOSUS. SMALL HAIRY WOOD-RUSH.

JUNCUS *Lin Gen. Pl.* HEXANDRIA MONOGYNIA.

Cal. 6-phyllus, *Cor.* o. *Capf.* 1-locularis.

Raii Syn. Gen. 27. HERBÆ GRAMINIFOLIÆ FLORE IMPERFECTO CULMIFERÆ.

JUNCUS *pilofus* foliis planis pilofis, corymbo ramofo. *Lin. Syst. Vegetab. p.* 280. *Sp. Pl.* 468. *Fl. Suec.* 308.

JUNCUS foliis planis, hirfutus, floribus umbellatis, folitariis, petiolatis, ariftatis. *Haller hist.* n. 1325.

JUNCUS *pilofus. Scopoli Fl. Carn.* n. 435.

GRAMEN nemorofum hirfutum latifolium minus. *Bauhin pin.* 7.

GRAMEN nemorofum hirfutum. *Ger. emac.* 19. majus *Park.* 1184

GRAMEN nemorofum hirfutum vulgare. *Raii Syn. p.* 416. Small hairy Wood-Rush. *Hudfon. Fl. Angl. p.* 151. *Lightfoot. Fl. Scot. p.* 186.

RADIX perennis, fibrofa, fibris numerofis, fufcis, ftolonibus brevibus acutis quoque inftruitur, ita ut fubrepens dici poteft.

ROOT perennial, and fibrous, fibres numerous and brown, it is alfo furnifhed with fhort pointed fhoots, fo that it may be called fomewhat creeping.

CULMI plures, ex eadem radice, fpithamæi et ultra, fuberecti, foliofi, fuperne nudi, fimplices, leves, ftriati, teretes, tribus aut quatuor geniculis minute protuberantibus inftructi.

STALKS many from the fame root, about a fpan in length, fometimes more, nearly upright, leafy, naked above, fimple, fmooth, ftriated, round, furnifhed with three or four joints, which do not protuberate.

FOLIA radicalia plurima, tres quatuorve uncias longa, lineas tres, trefque cum dimidiâ lata, ad bafin paulo anguftiora, parum concava, fuperne obfcure plerumque virentia et lævia glabraque, inferne dilutius virentia et glabra, ad margines autem, raris et longis pilis villofa, denfius autem hirfuta funt verfus eorum origines, fæpe rubentia, apice obtufiufcula et fubtruncata, caulina plana.

LEAVES next the root numerous, three or four inches long, and three lines or three and a half broad, fomewhat narroweft at the bafe, a little concave, above generally of a dull green colour, fmooth and rather glofsy, beneath of a paler green, and flightly glofsy, at the edges efpecially, covered with a few long hairs, which are moft numerous towards the bafe of the leaf, often of a reddifh colour, a little blunt and as it were cut off at the point, the ftalk leaves flat.

FLORES paniculati, panicula diffufa.

FLOWERS forming a fpreading panicle.

PEDUNCULI inæquales, pauci fimplices, plures proliferi, dichotomi et trichotomi, demum retro porrecti, omnes uniflori, flofculis intermediis fefsilibus.

FLOWER-STALKS of unequal lengths, a few of them fimple, moft of them proliferous, dichotomous or trichotomous, finally ftretcht out backward, all of them fupporting a fingle flower, the intermediate ones fefsile.

CALYX *Gluma* bivalvis, *fig.* 1. *Perianthium* hexaphyllum, foliolis oblongis, acuminatis, carinatis, concavis, ex purpureo fufcis, perfiftentibus, *fig.* 2. auct.

CALYX: a *Glume* of two valves, *fig.* 1. a *Perianthium* of fix leaves, which are oblong, pointed, keel'd, concave, of a purplifh brown colour and permanent, *fig.* 2. magnified.

COROLLA nulla.

COROLLA wanting.

STAMINA: FILAMENTA fex, capillaria, breviffima, ANTHERÆ oblongæ, erectæ, flavæ, *fig.* 3.

STAMINA: fix FILAMENTS, capillary and very fhort; ANTHERÆ oblong, upright, and yellow, *fig.* 3.

PISTILLUM: GERMEN triquetrum, acuminatum; STYLUS brevis, filiformis; STIGMATA tria, longa, filiformia, villofa, *fig.* 4.

PISTILLUM: GERMEN three-cornered, pointed; STYLE fhort, filiform: STIGMATA three, long, filiform, and villous, *fig.* 4.

The *Juncus pilofus, fylvaticus,* and *campeftris,* are diftinguifhed from the other fpecies, by their grafs-like hairy leaves; the firft of thefe has fome little affinity with the *campeftris* already figured, but differs from it, not only in its place of growth, but in having its flowers ftand fingly, and not in clufters; while the *campeftris* delights in expofed, the *pilofus* is found only in woods, and fhady fituations; and from this circumftance we may perhaps in fome degree account for its flowering earlier than any of the others, for if the feafon be not very unfavourable, it will begin to flower in February, and is ufually out of bloom the beginning of May.

We know of no ufe to which this fpecies, or the *fylvaticus,* is applicable: nor yet from the places they inhabit, can they be confidered in any degree noxious in Agriculture.

Juncus *pilosus*

Juncus sylvaticus.

Juncus Sylvaticus. Great Hairy Wood-Rush.

JUNCUS *Lin. Gen. Pl.* Hexandria Monogynia.

Cal. 6-phyllus. *Cor.* o. *Caps.* 1-locularis.

Raii Syn. Gen. 27. Herbæ graminifoliæ flore imperfecto culmiferæ.

JUNCUS *sylvaticus* foliis planis pilosis, corymbo decomposito, floribus fasciculatis sessilibus. *Hudson Fl. Angl. p.* 151.

JUNCUS foliis planis hirsutis, floribus paniculatis, fasciculatis *Haller hist.* n. 1324.

GRAMEN nemorosum hirsutum latifolium majus. *Scheuch. Agrost. p.* 317. *C. B. Pin.* 7.

GRAMEN nemorosum hirsutum latifolium maximum. *Raii Syn. p.* 416. The greatest broad-leaved hairy Wood-Grass.

GRAMEN luzulæ maximum. *J. B. II.* 493. *Lightfoot Fl. Scot. p.* 180.

Authors have contributed not a little to mislead students, by describing this species of Juncus, as uncommonly large and scarce, and it is probable that Mr. Ray would not have considered it as a species, had he not by accident met with some very luxuriant specimens of it; in certain situations it doubtless may be found very large, and tall, but it more usually occurs with a stalk a little more than a foot high; of some plants growing in my garden, close to each other, in a moist, but not very shady situation, the comparative height of the *Juncus campestris*, *pilosus*, and *sylvaticus*, was as follows, *campestris* 9 inches, *pilosus* 11, and *sylvaticus* 15; the account of its being a scarce plant is still more erroneous, as there is hardly a wood in the neighbourhood of London, nor as far as we have observed in any part of the kingdom, in which they do not grow plentifully together; they do so at least in Bishop's-Wood, Hampstead, which is near the spot where Mr. Ray describes his plant as growing.

By Linnæus this plant is considered as a variety only of the *pilosus*. Mr. Hudson and Baron Haller, examining it with more attention than Linnæus, make a distinct species of it, and give such a description of it as cannot fail to make it known.

To the characters given in their synonyms above quoted, we may add that the leaves are not only much broader, and more concave, but more sharply pointed than those of the *pilosus*, that it flowers three weeks or a month later, and that when the flowering is over, the flower-stalks of the *pilosus* are more reflexed or pendulous than those of the *sylvaticus*.

This species flowers in May, or earlier if the season be a mild one.

Alisma Plantago aquatica

ALISMA PLANTAGO. GREAT WATER-PLANTAIN.

ALISMA *Lin. Gen. Pl.* HEXANDRIA POLYGYNIA.

Cal. 3-phyllus. Petala 3. Sem. plura.

Raii Syn. Gen. 15. HERBÆ SEMINE NUDO POLYSPERMÆ.

ALISMA *Plantago foliis ovatis acutis, fructibus obtuse trigonis.* Lin. Syst. Vegetab. p. 288. Spec. Pl. p. 486. Fl. Suec. n. 323.

DAMASONIUM foliis ellipticis, lanceolatis, capitulo rotunde triquetro. Haller. Hist. n. 1184.

ALISMA *Plantago.* Scopoli Fl. Carn. n. 449.

PLANTAGO aquatica latifolia. Bauh. Pin. 190.

PLANTAGO aquatica major. Ger. emac. 417. Park. 1245. Raii Syn. 257. Great Water-Plantain. Hudson. Fl. Angl. ed. 2. p. 159. Lightfoot Fl. Scot. p. 193.

RADIX perennis, alba, bulbiformis, *radicata, densioribus* fibris capillata.

FOLIA omnia radicalia, longe petiolata, ovata, acuta, glabra, nervosa, integerrima, erecta, *subundulata, petiolis semiteretibus, basi vaginantibus, purpurascentibus.*

SCAPUS obtuse trigonus, nudus, lævis, pedalis ad tripedalem.

RAMI floriferi verticillatim circa scapum dispositi, atque *stellatim* circa ramos, numero quam *maxime varianter,* nudi.

STIPULÆ ad basin cujusvis verticilli, membranaceæ, *marcidæ,* vaginantes.

CALYX: PERIANTHIUM triphyllum, foliolis ovatis, aquisfoculis, concavis, *lineatis, patentibus,* margine membranaceis, *fig.* 1.

COROLLA: PETALA tria, suborotunda, purpurea, erecta, plana, *patentia, concaviuscula,* unguibus flavis, *fig.* 2.

STAMINA: FILAMENTA sex, *setacea,* subincurvata. ANTHERÆ virescentes. *fig.* 3.

PISTILLUM: GERMINA plurima, 12 et ultra, in orbem posita. STYLI tot quot germina, filiformes, erecti. STIGMATA simplicia, *fig.* 4. Pistillum auct. *fig.* 5.

ROOT perennial, white, somewhat bulbous, coated, and furnished with a tuft of numerous fibres.

LEAVES all springing from the root, standing on long *footstalks,* ovate, pointed, smooth, ribbed, perfectly entire, upright, slightly waved, the footstalks semicylindrical, at bottom sheathing and purplish.

STALK obtusely three-cornered, naked, smooth, from one to three feet in height.

BRANCHES producing the flowers disposed in whorls round the stalk and the little branches in a *similar* manner round them, varying greatly in number, and naked.

STIPULÆ at the base of each whirl, membranous, withered and sheathing.

CALYX: a PERIANTHIUM of three leaves, the leaves ovate, a little pointed, concave, marked with lines, spreading, membranous on the edge, *fig.* 1.

COROLLA: three PETALS, roundish, purple, upright on the edge, flat, spreading, concave-concave from each other, claws yellow, *fig.* 2.

STAMINA: six FILAMENTS, fine and tapering, slightly bending inwards. ANTHERÆ greenish. *fig.* 3.

PISTILLUM: GERMINA numerous, 12 or more placed in a circle. STYLES as numerous as the germina, threads, upright. STIGMATA simple, *fig.* 4. The Pistillum magnified, *fig.* 5.

The ancient Botanists, taken with the first appearance of things, and observing a similarity in the leaves of this plant to those of Plantain, without consulting the flower or fruit, made it at once a Plantago, though its fructification bears not the most distant affinity to that genus.

Baron HALLER observes, that in its acrimonious quality it comes near to the Crowfoots, and on the authority of FABRICIUS relates, that it has proved fatal to kine and other animals who have eaten it. From these effects he very properly queries how comes it to be considered by FLOYER as a *cooler* and astringent, and by BOCCONE as useful in the Piles.

Externally applied it blisters; taken internally it produces the same effect as the Crowfoots. Cattle are much injured, and sometimes killed by it. Atrophy and immobility of the hind parts of the body are the effects of which it is productive. LINDESTOLPIUS, *Bergman's Dissertatio Lucrosæ sunt Plantæ inutiles,* &c. 1783.

There is no plant more common than this species of Water Plantain in and by the sides of ponds, rivers, &c. It flowers in July, August, and September.

Alisma Damasonium

ALISMA DAMASONIUM. STARRY-HEADED WATER-PLANTAIN.

ALISMA *Lin. Gen. Pl.* HEXANDRIA POLYGYNIA.

Cal. 3-phyllus. *Petala* 3. *Sem.* plura.

Raii Syn. Gen. 27. HERBÆ MULTISILIQUÆ SEU CORNICULATÆ.

ALISMA *Damasonium* foliis cordato oblongis, floribus hexagynis, capsulis subulatis. *Lin. Syst. Vegetab. p.* 350. *Sp. Pl. p.* 486.

PLANTAGO aquatica stellata. *Bauh. Pin.* 190.

DAMASONIUM stellatum Dalechampii. *I. B.* III. 789.

PLANTAGO aquatica minor stellata. *Ger. emac.* 417.

PLANTAGO aquatica minor muricata. *Park.* 1245. *Raii Syn.* Star-headed Water-Plantain. *Hudf. Fl. Angl. ed.* 2. *p.* 158.

RADIX perennis, fibrosa, fibris plurimis, densissime capillatis, implexisculis, ex fusco-aurantiacis, in limum profunde demissis, junioribus albis.

FOLIA longe petiolata, natantia, cordato-oblonga, integerrima, utrinque glabra, obtusa, margine ipsa perpusillorum, subtus nervosa, nervis duobus vix procubentibus parallelis prope marginem.

PETIOLI obtuse trigoni, subdiaphani, spongiosi, ad basin lati, et membrana albida utrinque instructi.

SCAPUS spithamæus, teres, lævis, nudus, crassiusculus, superne sordide purpureus, multiflorus.

FLORES albi, subumbellati.

UMBELLÆ plerumque tres, inferior lateralis, octoradiata, proxima superior fexradiata, suprema triradiata, numerus vero variat in diversis plantis.

INVOLUCRUM umbellæ triphyllum, foliolis ovatolanceolatis, membranaceis, marcescentibus.

PEDUNCULI qui radii umbellæ, teretes, nudi, sesquiunciales, superioribus brevioribus.

CALYX: PERIANTHIUM triphyllum, foliolis subovatis, obtusis, concavis, patentibus, apice membranaceis, cito marcescentibus, *fig.* 1.

COROLLA: PETALA tria, subrotunda, alba, tenera, ungue flexo, *fig.* 2.

STAMINA: FILAMENTA sex, subulata, flavescentia, corolla breviora: ANTHERÆ oblongæ, flavæ, *fig.* 3.

PISTILLUM: GERMINA plerumque sex, subulata, erecta: STYLI nulli: STIGMATA villosa, inflexiflexa, *fig.* 4.

PERICARPIUM: CAPSULÆ sex, patentes, subulatæ, inferne compressæ, uniloculares, monospermæ vel dispermæ, *fig.* 5.

SEMEN oblongum, obtusum, nigricans, nitidum, ad lentem punctis exasperatum, sulco per medium utrinque longitudinali, *fig.* 6.

ROOT perennial, fibrous, fibres numerous, thickly matted together, mostly simple, of a brownish orange colour, striking deeply into the mud, the young ones white.

LEAVES standing on long footstalks, swimming, of an oblong heart shape, perfectly entire, smooth on both sides, obtuse, the very edge purplish, ribb'd on the under side, two very slightly prominences, parallel ribs near the margin.

LEAF-STALKS obtusely three-cornered, somewhat transparent, spongy, broad at the base, and edged on each side with a whitish membrane.

STALK about a span long, round, smooth, naked, clumsy, of a dirty purple colour above, many-flower'd.

FLOWERS white, growing umbel like.

UMBELS for the most part three, the lowermost lateral, eight-rayed, the next above six-rayed, the uppermost three-rayed, the number however varies in different plants.

INVOLUCRUM of the umbel three-leav'd, leaves ovato-lanceolate, membranous, and withering.

FLOWER-STALKS which form the rays of the umbel, round, naked, an inch and a half in length, the upper ones shortest.

CALYX: a PERIANTHIUM of three leaves, the leaflets nearly ovate, obtuse, concave, spreading, membranous at the top, and soon withering, *fig.* 1.

COROLLA composed of three roundish, white, tender PETALS with yellow claws, *fig.* 2.

STAMINA: six tapering yellowish FILAMENTS, shorter than the corolla: ANTHERÆ oblong and yellow, *fig.* 3.

PISTILLUM: GERMINA for the most part six in number, tapering, upright: STYLES none: STIGMATA villous, somewhat reflexed, *fig.* 4.

SEED-VESSEL: six spreading CAPSULES, tapering to a point, flattened below, one-cell'd, a single seed or two in each, *fig.* 5.

SEED oblong, obtuse, blackish, shining, when magnified appearing rough with little prominent points, a groove running down the middle on each side, *fig.* 6.

Not very uncommon in the neighbourhood of London, in ditches, stagnant waters, and ponds, especially such as have been formed by the digging of gravel: particularly plentiful in such like ponds on Wandsworth Common, with *Sparganium simplex* : also, about Clapham, Walworth, &c.

Flowers from June to September.

It is not remarkable for its qualities or uses.

TOURNEFORT makes a distinct genus of the *Damasonium*, referring the *Alisma Plantago* and *ranunculoides* to the genus *Ranunculus*.

RAY also separates it from the *Plantago aquatica*, but observes that it agrees with it in its tripetalous flowers, though it differs in its seed-vessels.

Notwithstanding this discrepance in the seed-vessels, the other parts of its fructification, joined to its general habit, in our humble opinion, fully justify LINNÆUS in making it an *Alisma*.

RUMEX ACETOSELLA. SHEEP'S SORREL.

RUMEX *Lin. Gen. Pl.* HEXANDRIA TRIGYNIA.

    Cal. 3-phyllus. Petala 3, conniventia. Sem. 1. triquetrum.

   Raii Syn. Gen. 5. Herbæ flore imperfecto seu flamineo (vel apetalo potius).

RUMEX *Acetofella floribus dioicis foliis lanceolato-haltatis. Linn. Syst. Vegetab. p. 286. Sp. Pl. 481. Fl. Suec. n. 319.*

LAPATHUM *fexubus feparatis, foliis fagittatis, hamis acutis recurvis. Haller hist. 1596.*

LAPATHUM *Acetofella. Scopuli Fl. Carn. n. 439.*

ACETOSA *arvenfis lanceolata. Bauhin. Pin. p. 114.*

OXALIS *tenuifolia. Ger. emac. 397.*

ACETOSA *minor lanceolata. Parking. 744.*

LAPATHUM *acetofum repens lanceolatum. Raii Syn. p. 143.* Sheep's Sorrel. *Hudfon Fl. Angl. p. 156. Lightfoot Fl. Scot. p. 191.*

RADIX perennis, fublignofa, repens, fufca.	**ROOT** perennial, of a brown colour, fomewhat woody, and creeping.
CAULIS palmaris ad pedalem, erectus, lævis, ftriatus, fubangulofus, ramofus.	**STALK** from a hand's breadth to a foot in height, upright, fmooth, ftriated, fomewhat angular, branched.
FOLIA alterna, petiolata, inferiora lanceolato-haftata, hamis furfum recurvis, in umbrofis fubglauca, in apricis ut et tota planta fanguinea, fuperiora lineari-lanceolata.	**LEAVES** alternate, ftanding on foot-ftalks, the lower ones lanceolate, and halbert-fhaped, the lobes forming the halbert, ufually bent upwards, in fhady fituations fomewhat glaucous, in expofed ones of a blood colour, as well as the whole plant, the upper ones entire, betwixt linear and lance-fhaped.
PETIOLUS longitudine folii, inferne ftriatus, fuperne canaliculatus, bafi vaginans, vaginæ apice membranacei, albi, lacerâ, fæpe reflexâ.	**LEAF-STALK** the length of the leaf, on the under fide ftriated, above fingle-channeled, forming a fheath at bottom, the tip of which is membranous, white, torn, and often reflexed.
SPICÆ plurimæ, nudæ, fubramofæ, fæpe nutantes.	**SPIKES** numerous, naked, fomewhat branched, and often drooping.
FLORES mafculi et feminei in diftinctis plantis, minimi; *fig. 1,* 1. flos mafculus auctus; *fig. 3.* femineus; *fig. 4.* femen magnitudine naturali, *fig. 5.* idem auctus.	**FLOWERS** male and female in feparate plants, very minute; *fig. 1,* 2. a male flower magnified; *fig. 3.* a female flower; *fig. 4.* the feed of its natural fize; *fig. 5.* the fame magnified.

In reprefenting the two fexes (which occur in this as well as in the common Sorrel) we have intended that one of them fhould exprefs the plant in its dwarf ftate, as it ufually occurs on very dry, hilly paftures. In fuch fituations the whole plant is frequently found of a bright red colour. In more fhady afpects it grows taller, and the leaves affume a greener hue. Wherever it abounds we may in general look on it as a fure indication of a **dry**, barren foil. HALLER obferves, that it is often found growing in Coal-yards (*arvis carbonariorum*).

Agriculturally confidered, we muft number it with the **weeds**, and with thofe too, from its creeping roots, of difficult extirpation.

It is found in flower from June to September.

Rumex Acetosella

ERICA VULGARIS. COMMON HEATH.

ERICA *Lin. Gen. Pl.* OCTANDRIA MONOGYNIA.
    Cal. 4-phyllus. Cor. 4-fida. *Filamenta* receptaculo inferta. *Antheræ* bifidæ.
    Caps. 4-locularis.

    Raii Syn. ARBORES ET FRUTICES.

ERICA *vulgaris* antheris ariftatis, corollis campanulatis fubæqualibus, calycibus duplicatis, **foliis oppofiti-**
    fagittatis. *Lin. Syft. Vegetab.* p. 321. *Sp. Pl.* p. 501. *Fl. Suec.* n. 336.
ERICA foliis leviø adpreffis fimplicibus, floribulus calicatis. *Haller. Hift.* n. p. 1012.
ERICA *vulgaris. Scopoli Fl. Carn.* n. 460.
ERICA *vulgaris glabra. Baub.·Pin.* 485.
ERICA *vulgaris feu pumila. Ger. emac.* 1380.
ERICA *vulgaris. Parkinf.* 1480. *Raii Syn.* 470. **Common Heath or Ling.** *Scot.* Hather. *Hudfon.*
    Fl. Angl. ed. 2. p. 165. *Lightfoot Fl. Scot.* p. 204.

Fruticulus pedalis, bipedalis et ultra, valde ramofus, ramis fub-erectis, teretes, pubefcentes, rubicundi.	A fmall fhrub, a foot or two in height, or more, very much branched, the branches moftly upright, round, downy, and reddifh.
FOLIA oppofita, circa ramulos in quatuor feries imbricata, feffilia, fagittata.	LEAVES oppofite, feffile and arrow-fhaped, placed round the fmall branches in four rows.
FLORES purpurei, fpicati, fubfecundi.	FLOWERS purple, growing in a fpike, moftly all one way.
PEDUNCULI breviffimi, longitudine foliorum.	FLOWER-STALKS very fhort, the length of the leaves.
CALYX: duplex, perfiftens, exterior breviffimus, tetraphyllus, foliolis ovatis, acutis, patentibus, e viridi purpurafcentibus, ad latera ciliaris, interior cum corolla concolor, tetraphyllus, foliolis ovato-lanceolatis, nitidis, corolla longioribus, demum inflexis, *fig.* 1, 2.	CALYX: double, and permanent, the outermoft very fhort, compofed of four leaves, which are ovate, pointed, fpreading, partly green, and partly purple, when magnified hairy on the edges, the inner one the fame colour as the corolla, compofed of four fomewhat lanceolate leaves, fhining, longer than the corolla, finally bending inward, *fig.* 1, 2.
COROLLA monopetala, purpurea, quadripartita, corolla brevior, inclufa, *fig.* 3.	COROLLA monopetalous, purple, deeply divided into four fegments, fhorter than the corolla, and inclofed within it, *fig.* 3.
STAMINA: FILAMENTA octo, alba. ANTHERÆ fub-coadunatæ, auranthiæ, bicornes, *fig.* 4, 5.	STAMINA: eight white FILAMENTS. ANTHERÆ fomewhat united, orange-coloured, each furnifhed with two little horns, *fig.* 4, 5.
PISTILLUM: Germen villofum. Stylus calyce longior, fortium curvatus. Stigma quadrifidum, *fig.* 6.	PISTILLUM: Germen villofus. Style longer than the calyx, bent upward. Stigma quadrifid, *fig.* 6.

There is, perhaps, no tribe of plants whofe flowers affume a greater variety of form than thofe of the prefent genus. Such as have had opportunities of examining many of the foreign heaths, muft affent to the truth of this obfervation; and fuch as have not, need only confult the prefent fpecies, and compare the diffections with thofe of the *Erica cinerea*, and *Tetralix* already figured, to be perfectly convinced of it; fo great indeed has this difference appeared to fome botanifts, that they have divided them into diftinct genera.

Africa produces more heaths than the whole world befides. Next to Africa, Europe is the moft productive; and almoft every part of this quarter of the globe, efpecially the northern, abounds with this fpecies. Linnæus remarks, in his *Flora Lapponica*, that, in fome of the diftricts through which he paffed, fcarce any plant was to be feen but the barren heath, which every where covered the ground, and could no ways be extirpated. The country people, he obferves, had no idea that there were two plants which would finally overfpread and deftroy the whole earth, viz. Heath and Tobacco.

Exclufive of the admiration which the bloffoms of this fpecies in particular impart to our dreary waftes at the clofe of fummer, it anfwers many important purpofes in natural as well as rural œconomy.

While its branches afford fhelter to many of the feathered tribe, its feeds form a principal part of their food, efpecially thofe of the Groufe kind; and here we may remark a particular provifion of nature in forming the feedveffel, &c. in fuch a manner as to preferve the feeds a whole year, or longer, whence they have a conftant fupply. The foliage of this fpecies affords nourifhment to the caterpillar of the *Phalæna quercus Linnei*, or great Egger Moth; we obferved many inftances of this in our northern tour. Bees are well known to collect largely from the bloffoms of heath; but fuch honey is browner, coarfer, and of lefs value than fuch as is collected where no heath grows. According to Linnæus's experiments, no kind of cattle appear to be fond of it. Horfes and Oxen will eat it; fheep and Goats fometimes eat, fometimes reject it. Cattle, not accuftomed to browfe on heath, give bloody milk; but are foon cured, by drinking plentifully of water, *Peasant's Tour,* p. 219.

Heath or Hather is applied to many œconomical purpofes among the Highlanders: they frequently cover their houfes with it inftead of thatch, or elfe twift it into ropes, and bind down the thatch with them in a kind of latthice work. In moft of the weftern ifles they dye their yarn of a yellow colour, by boiling it in water with the green tops and flowers of this plant. In Rum, Skye, and the Long Ifland, they frequently tan their leather in a ftrong decoction of it. Formerly the young tops are faid to have been ufed alone to brew a kind of ale; and even now, I was informed, that the inhabitants of Ifa and Jura ftill continue to brew a very potable liquor, by mixing two-thirds of the tops of Hather, and one-third of malt. This is not the only refrefhment that Hather affords; the hardy Highlanders frequently make their beds with it, laying the roots downwards, and the tops upwards, which, though not quite fo foft and luxurious as beds of down, are altogether as refrefhing to thofe who are fo inured, and perhaps much more healthy. *Lightfoot Fl. Scot.* p. 205.

In moft parts of Great Britain, Heath is in general ufe for making brooms; and for this purpofe is ufually cut when in bloffom. The turf, with the Heath growing on it, is cut up, dried, and ufed for fuel by the poor cottager. It is alfo in ufe for heating ovens, for mending bad roads where better materials are wanting, and for making drains under-ground.

This fpecies, as well as the others, is fometimes found with white bloffoms, and a variety with hoary leaves is not uncommon, particularly on Bagfhot Heath. Some authors have improperly confidered this as the *Erica ciliaris* of Linnæus.

The Dodder very frequently entwines itfelf about this plant, and gives it an appearance which may puzzle, if not miflead, the inexperienced botanift.

Erica vulgaris

Spergula arvensis.

SPERGULA ARVENSIS. CORN SPURREY.

SPERGULA *Linnæi Gen. Pl.* DECANDRIA PENTAGYNIA.
 Raii Syn. Gen. 24. HERBÆ PENTAPETALÆ VASCULIFERÆ.
SPERGULA *arvensis foliis verticillatis, floribus decandris. Linn. Syst. Vegetab. p.* 363. *Sp. Pl. p.* 630.
 Fl. Suec. n. 429.
ALSINE *foliis verticillatis, seminibus rotundis. Haller. hist. n.* 879.
ALSINE *spergula dicta major. Bauhin. Pin.* 251.
SAGINA *spergula. Ger. emac.* 1125.
SAGINA *Spergula major. Park. hist.* 561. *Raii Syn. p.* 351. *Spurrey. Hudson. Fl. Angl. ed.* 2. *p.* 203.
 Lightfoot Fl. Scot. p. 242.

RADIX annua, fibrosa.	ROOT annual and fibrous.
CAULES plures, fpithamæi, feu pedales, fuberecti, teretes, læves, fuperne vifcofi, geniculis globofis.	STEMS numerous, about a fpan or a foot in length, nearly upright, round, fmooth, on the upper part clammy, joints globular.
STIPULÆ ad geniculis binæ, breviffimæ, apicibus inferioribus reflexis.	STIPULÆ growing in pairs at the joints, very fhort, the tips of the lower ones reflexed.
FOLIA verticillata, fafciculos duos conftituentia, foliolis octo cinéter in quovis fafciculo, interioribus fenfim minoribus, linearis, teretia, apicibus flavis, dorfo lineâ exaratis, fuperioribus vifcofis.	LEAVES growing in whorls, and forming two bundles, about eight in each bundle, the inner ones gradually fmallifh, linear, round, tips yellow, with a deep furrow on the back, the upper ones clammy.
FLORES albi, pulchelli, paniculati, paniculâ dichotoma.	FLOWERS white, pretty, growing in a panicle, which is dichotomous.
PEDUNCULI vifcofi, peractâ florefcentiâ penduli.	PEDUNCLES clammy, hanging down when the flowering is over.
CALYX: PERIANTHIUM pentaphyllum, foliolis ovatis, obtufiufculis, concavis, patentibus, perfiftentibus, marginibus albidis, *fig.* 1.	CALYX: a PERIANTHIUM of five leaves, the leaves ovate, bluntifh, concave, fpreading, permanent, the edges whitifh, *fig.* 1.
COROLLA: PETALA quinque, ovata, acutiufcula, concava, calyce longiora, ungue brevi affixa, *fig.* 2.	COROLLA: five PETALS, ovate, a little pointed, concave, longer than the calyx, affixed by a fhort claw, *fig.* 2.
STAMINA: FILAMENTA decem, alba, fubulata, ANTHERÆ fubrotundæ, flavæ, *fig.* 3.	STAMINA: ten FILAMENTS, white, tapering; ANTHERÆ roundifh and yellow, *fig.* 3.
PISTILLUM: GERMEN fubrotundum; STYLI quinque, breves, reflexi; STIGMATA fimplicia, *fig.* 4.	PISTILLUM: GERMEN roundifh; STYLES five, fhort, reflexed; STIGMATA fimple, *fig.* 4.
PERICARPIUM: CAPSULA ovata, tecta, unilocularis, quinquevalvis, *fig.* 5.	SEED-VESSEL: an ovate CAPSULE covered, by the remaining calyx, of one cavity and five valves, *fig.* 5.
SEMINA plurima, majufcula, nigricantia, depreffo-globofa, punctulis rufis punctiuolis ad lentem exafperata, annulo manifefto cincta, *fig.* 6, 7.	SEEDS numerous, rather large, blackifh, round, with a fmall degree of flatnefs, if viewed with a magnifier befet with fmall, reddifh, prominent points, and encircled with a manifeft ring, *fig.* 6, 7.

The *Spergula arvensis* is feldom found but in a fandy foil; and as that kind of foil does not abound much in the neighbourhood of London, to this fpecies of Spergula may be confidered as one of our fcarcer robots. On fome parts of Hampftead-Heath, and in the neighbourhood of the Spaniard, we have often noticed it, as well as in the footpaths at Charlton. In fome fandy fields near Carfhalton, in Surrey, we have feen it fo plentiful as to appear like the intended crop. As no ufe is made of it with us, it may be confidered as one of the word weeds to which a fandy foil is fubject. Abroad, however, it is an object of cultivation. In fome parts of Flanders, Germany, and Norway, they feed their cattle with the plant, and their poultry with its feeds; but as Tares and Buck-wheat, which are far more productive, as well as nutritious, may be cultivated in a fimilar foil, our Farmers do wifely in rejecting it.

It is found in bloffom from July to September.

We have not found this plant unufually fubject to vary in the number of its ftamina; nor have we obferved it to vary fo much in any other refpect as to make us fufpect we had here the *Spergula pentandra* of Linnæus, which Mr. Hudson makes a variety of the *arvensis*, contrary to the opinion of fome of the greateft authorities. If the difference betwixt thefe two plants was to depend folely on the number of its ftamina, we fhould be extremely ready to confider them as the fame; but Ray, whofe opinion muft be allowed to have great weight, defcribes the *pentandra* as a fpecies totally diftinct from the *arvensis*. He does not found his fpecific difference on the number of its ftamina; but on charchters, lefs fubject to variation; the leaves at the joints, he obferves, are fewer and thicker, the plant flowers early, and foon goes off (investing of which takes place in the *arvensis*;) and adds, that Dr. Sherard was obferved it in fandy places in Ireland.

To them that other Authors have likewife entertained an opinion of its being a diftinct fpecies, we fhall quote their refpective fynonyms.

Spergula foliis filiformibus verticillatis varie feminibus nigris. *Sauv. Monfp.* 167.
Alfine fpergulæ facie minus feminibus emarginatis. *Tourn. inst.* 244. *Vaill. Paris* 8.
Alfine fpergulæ facie minima. *Mag. Monfp.* 22.
Arenaria tenuifolia cerea, flore albo, femine fufco follieato cincto. *Rupp. Jen.* 101.
Spergula minor, femine foliaceo nigro circulo rotundiore in albo cincto. *Morif. hist.* 2. *p.* 551. *Hof.* 28. *Dill Gif.* 48. *E. N. C. cent.* 3 *p.* 375. *t.* 4.

On thefe feveral teftimonies we cannot but conclude, that there exifts fuch a plant as the *Pentandra*; and can we avoid expreffing a wifh, that fome gentleman, whofe refidence may afford him an opportunity of obferving its hiftory, will favour us with a more compleat account of it.

Agrimonia Eupatoria

AGRIMONIA EUPATORIA. AGRIMONY.

AGRIMONIA *Lin. Gen. Pl.* DODECANDRIA DIGYNIA.

Cal. 5 dentatus, altero obvallatus.   Petala 5.   Sem. 2, in fundo calycis.

Raii Syn. Gen. 10. HERBÆ FLORE PERFECTO SIMPLICI SEMINIBUS NUDIS SOLITARIIS SEU AD SINGULOS FLORES SINGULIS.

AGRIMONIA *Eupatoria* foliis caulinis pinnatis: impari petiolato, fructibus hispidis.   *Lin. Syst. Veg. p.* 372.   *Sp. Pl. p.* 643.   *Fl. Suec. n.* 423.

AGRIMONIA foliis pinnatis, pinnulis alterne minimis.   *Haller Hist.* 591.

AGRIMONIA *Eupatoria. Stapeli Fl. Carn. n.* 567.

EUPATORIUM veterum feu Agrimonia. *Bauh. Pin.* 321.

AGRIMONIA *Ger. emac.* 712.

AGRIMONIA vulgaris.   *Park.* 594.   *Raii Syn. p.* 202.   Agrimony.   *Hudfs. Fl. Angl. ed.* 2. *p.* 206.   *Lightfoot Fl. Scot. p.* 247.

RADIX perennis, ramofa, rubefcens, fquamis nigricantibus obfefa.

CAULIS pedalis ad tripedalem, erectus, teres, obfolete angulofus, hirfutus, rubicundus aut rubro punctatus, fimplex vel ramofus.

FOLIA alterna, fubambrofinea, hirfuta, interrupte pinnata cum impari, 5 vel 6 juga, pinnæ lobo oppofitæ, feffiles, fubovatæ, venofæ, hirtæ, ciliatæ, pinnulæ plerumque integræ aut trifidæ.

STIPULÆ duæ, oppofitæ, majufculæ, amplexicaules, patentes, profunde ferratæ.

BRACTEÆ trifidæ, laciniis linearibus, hirfutæ.

SPICA terminalis, elongata, hirfuta, floribus breviter pedicellatis.

CALYX: PERIANTHIUM monophyllum, quinquefidum, fuperum, perfiftens, laciniis ovatis, acutis, *fig.* 2. extus fetis filiformibus, rigidis, apicis purpureis, uncinatis, circlum, *fig.* 2. intus fubftantia flava glandulofa claufum: Involucrum ad bafin germinis diphyllum foliolis binis feu tridentatis, *fig.* 3.

COROLLA: PETALA quinque, fubovata, flava, patentia, feffilia, fubftantia glandulofa calycis inferta, *fig.* 4.

STAMINA: FILAMENTA undecim, feu duodecim, luteifcentia, curvata, cum petalis inferta. ANTHERÆ didymæ, compreffæ, *fig.* 5.

PISTILLUM: GERMEN inferum, *fig.* 6. STYLI duo, curvati, longitudine ftaminum. STIGMATA obtufa, *fig.* 7.

PERICARPIUM: CAPSULA e calyce orta, nutans, extra fulcatum, fuperne cincta ariftis uncinatis, unilocularis, *fig.* 8.

SEMINA duo, fubrotunda, glabra, *fig.* 9.

ROOT perennial, branched, of a reddish colour, befet with blackish fcales.

STALK from one to three feet high, upright, round, faintly angular, hirfute, reddish or dotted with red, fingle or branched.

LEAVES alternate, fomewhat clammy, hirfute, interruptedly pinnated with an odd one at the end, compofed of five or fix pair of pinnæ, pinnæ ufually oppofite, feffile, fomewhat ovate, veiny, ferrated, edged with hairs, the fmall pinnæ for the moft part entire or trifid.

STIPULÆ two, oppofite, rather large, embracing the ftalk, fpreading, and deeply ferrated.

FLORAL-LEAVES trifid, the fegments linear and hairy.

SPIKE terminal, elongated, hirfute, the flowers ftanding on very fhort foot-ftalks.

CALYX: a PERIANTHIUM of one leaf, divided into five fegments, placed above the germen, and permanent, the fegments ovate, pointed, *fig.* 2. externally furrounded with rigid, filiform, hooked, briftles, purple at the points, *fig.* 2. within clofed with a yellow glandular fubftance: Involucrum at the bafe of the germen, compofed of two leaves, each of which has two or three teeth, *fig.* 3.

COROLLA: five PETALS, fomewhat ovate, yellow, fpreading, feffile, inferted into the glandular fubftance of the calyx, *fig.* 4.

STAMINA: eleven or twelve FILAMENTS, of a yellow difcolour, bent and inferted with the petals. ANTHERÆ compofed of two lobes and flattened, *fig.* 5.

PISTILLUM: GERMEN beneath the calyx, *fig.* 6. STYLES two, bent, the length of the ftamina. STIGMATA blunt, *fig.* 7.

SEED-VESSEL: a CAPSULE, arifing from the calyx, drooping, grooved on the outfide, on the upper part furrounded with hooked beards, of one cavity, *fig.* 8.

SEEDS two, roundish and fmooth, *fig.* 9.

Agrimony is a plant of very general growth, being found not only in Europe, but in Virginia and Japan.

It has been chiefly regarded as a medicinal plant, and as fuch is often raifed in gardens. Culture does not feem to produce any material change in its quality. Another fpecies or variety, of foreign original, common alfo in our gardens, and differing little in appearance from our indigenous Agrimony, pronifes to be fuperior to it in virtue, as its tafte is more aromatic, and its fmell much ftronger, and very agreeable. CASPAR BAUHINE calls it *Eupatorium odoratum*. FABIUS COLUMNA *Eupatorium Diofcoridis odoratum et montanum. Lowe Dict. p. Ait. p.* 29.

The leaves of Agrimony have a flightly bitterish, rough tafte, accompanied with an agreeable, though very weak, aromatic flavour. The flowers are in fmell ftronger, and eaten agreeable, than the leaves, and in tafte fomewhat weaker. They readily give out their virtues both to water and rectified fpirit. The leaves impart to the former a greenish yellow, to the latter a deep green colour: the flowers yield their own deep yellow tincture to both menftrua. *Id.*

Agrimony is one of the milder corroborants; and in this intention is fometimes employed, efpecially among the common people, againft habitual diarrhœas, and cachectic and other indifpofitions, from a lax ftate of the folids. Infufions of the leaves, which are not ungrateful, may be drank as tea. It is fometimes joined with other ingredients in diet drinks for purifying the blood, and in pectoral apozems. *Id.*

This plant delights in a dry foil, and grows almoft every where, in this kingdom, in open paftures, in the borders of fields, and by the fides of hedges and ditches, flowering from July to September.

Cattle in general diflike, and leave it untouched.

Spiræa Ulmaria. Meadow-Sweet.

SPIRÆA *Lin. Gen. Pl.* Icosandria Pentagynia.

Cal. 5-fidus. Petala 5. Cssf. polyspermæ.

Raii Syn. Gen. 15. Herbæ semine nudo polyspermæ.

SPIRÆA *Ulmaria* foliis pinnatis : impari majore lobato, floribus cymosis. *Lin. Syst. Vegetab. p.* 393. *Sp. Pl. p.* 702. *Fl. Suec. n.* 440.

FILIPENDULA foliis pinnatis, acute serratis, minimis intermissis, extrema trilobata maxima. *Haller. hist. n.* 1135.

SPIRÆA *Ulmaria Scopoli Fl. Carn.* n. 603.

BARBA CAPRI floribus compactis. *Bauh. Pin.* 164.

ULMARIA *J. B.* III. 488.

REGINA PRATI *Ger. emac. p.* 1043.

ULMARIA vulgaris. *Parkins.* 592. *Raii Syn. p.* 259. Meadow-Sweet. *Hudson Fl. Angl. ed.* 2. *p.* 217. *Lightfoot Fl. Scot. p.* 259.

RADIX perennis, crassitie minimi digiti, obliqua, rubicunda, fibris plurimis ex sessu lutescentibus descendentibus instructa.

ROOT perennial, the thickness of the little finger, oblique, reddish, furnished with numerous fibres of a brownish yellow colour, running deep into the earth.

CAULIS bi seu tripedalis et ultra, erectus, foliosus, angulatus, glaber, hinc inde rubicundus, plerumque simplex.

STALK from two to three feet high or more, upright, leafy, angular, smooth, here and there of a reddish colour, for the most part unbranched.

FOLIA alterna, petiolata, pinnata, 3-vel 5-juga : foliolis oppositis, sessilibus, ovato-oblongis, supra viridibus, glabris, lucidiusculis, lineatis, minutim venulosis, rugosis, subtus nervosis, minutim tomentosis, cinereis, margine inciso-dentatis, undique serratis, minutim ciliatis ; terminatis foliolo majore, trifido-palmato.

LEAVES alternate, standing on foot-stalks, pinnated, from three to five pair, opposite, sessile, ovato-oblong, above green, smooth and somewhat shining, minutely veined, and wrinkled, the veins impressed, beneath ribbed, covered with an ash-coloured downy substance, the edge jagged, serrated, and finely edged with hairs, the terminal pinna large and deeply divided into three segments.

PETIOLI subtus convexi, supra concavi ; radicales triplo longiores.

LEAF-STALKS convex beneath, concave above, those of the radical leaves three times as long as the others.

STIPULÆ amplexicaules, acutæ, margine undique serratæ, minutim ciliatæ ; partiales in petiolo communi intra singulum par pinnarum, sub oppositæ, parvæ, inæqualis magnitudine, ovatæ, dentato-serratæ, pariter subtus tomentosæ.

STIPULÆ stem-clasping, pointed, serrated, and finely edged with hairs, the partial ones on the common foot-stalk betwixt each pair of pinnæ, nearly opposite, small, unequal in size, ovate, indented or serrated, and like the pinnæ downy underneath.

CORYMBUS terminalis, erectus, minutim pubescens, pedunculatus, nudus, compositus e cymis pluribus inæqualibus, intermedia sessili.

CORYMBUS terminal, upright, slightly pubescent, stalked, naked, composed of several unequal cymæ, the intermediate one sessile.

CALYX : Perianthium monophyllum, subcampanulatum, sub lentem pubescens, pallidum, quinquefidum, laciniis ovatis, acutis, demum reflexis, *fig.* 1.

CALYX : a Perianthium of one leaf, somewhat bell-shaped, if magnified slightly downy, of a pale colour, divided into five segments, which are ovate, pointed, and finally reflexed, *fig.* 1.

COROLLA : Petala quinque, albida, oblongo-rotundata, unguiculata, patentia, calyce duplo longiora, *fig.* 2.

COROLLA : five whitish Petals, oblong, roundish, clawed, spreading twice the length of the calyx, *fig.* 2.

STAMINA : Filamenta viginti plura, filiformia, flavescentia, longitudine corollæ, calyci inserta. Antheræ subrotundæ, flavescentes, *fig.* 3.

STAMINA : twenty Filaments or more, filiform, yellowish, the length of the corolla, inserted into the calyx. Antheræ nearly round, and yellowish, *fig.* 3.

PISTILLUM : Germina quinque, sex, sive plura, Styli totidem, superne incrassati, reflexa ; Stigmata capitata, *fig.* 4.

PISTILLUM : Germina five, six, or more ; Styles as many, thickened above and turned back ; Stigmata forming little heads, *fig.* 4.

PERICARPIUM : Capsulæ plurimæ, spiraliter contortæ, *fig.* 5.

SEED-VESSEL : Capsules several, twisted together spirally, *fig.* 5.

The Meadow-Sweet has been justly celebrated for its fragrance and beauty, the agreeable odour which the whole plant, but more particularly the flowers, diffuse, has recommended it for the purpose of scenting rooms, and purifying the air, by strewing it on the floors ; it is said not to affect the head like other perfumes : the leaves also, like those of Burnet, impart an agreeable flavour to wine and other liquors.

As an ornamental plant, it has long held a place in our gardens, not only in its wild state, but with variegated leaves and double flowers.

It puts in its claim also for medicinal virtues, which, however, do not appear to be of the most powerful kind ; the leaves are recommended as mildly astringent, and useful in Dysenteries ; the flowers are said to be antispasmodic and diuretic : their pleasant smell, in which their virtue resides, is soon dissipated by keeping.

It grows plentifully in wet meadows and by the sides of ponds and ditches, flowering from July to September.

Horses and kine are said to refuse it, sheep to eat it, and goats to be particularly fond of it ; as it forms a great part of the pasturage in some meadows, it is of consequence for the husbandman more clearly to ascertain whether horses and cows refuse the young foliage, and whether they reject the whole plant when made into hay.

We have frequently observed small red tubercles on the leaves, which we have supposed to be occasioned by some species of Cynips.

Spiraea Ulmaria.

ROSA CANINA. DOG ROSE.

ROSA *Lin. G.n. Pl.* ICOSANDRIA POLYGYNIA.

Cal. urceolatus, quinquefidus, carnosus, collo constrictus. *Petala* 5. *Sem. plurima*, hispida, calycis interiori lateri affixa.

Raii Syn. ARBORES ET FRUTICES

ROSA *caules geminalibus ovatis, pedunculisque glabris, caule petiolisque aculeatis. Lin. Syst. Vegetab. p. 394. Sp. Pl. p. 704. Fl. Suec. n. 441.*

ROSA *spinis aduncis, foliis septenis, calycibus tomentosis, segmentis pinnatis et semipinnatis, tubis brevissimis. Haller. Hist. n. 1601.*

ROSA *caninae. Scopoli Fl. Carn. n. 604.*

ROSA *sylvestris vulgaris flore odorato incarnato. Bauh. Pin. p. 483.*

ROSA *sylvestris inodora* L *canina. Park. p. 1017. sylvestris alba cum rubore folio glabro. I. B. II. p. 43. Raii Syn. p. 454.* Cynosbatos et Cynorrhodon Officinarum. The common wild Briar or Dog's Rose, the Hep tree. *Hudson. Fl. Angl. ed. 2. p. 220. Lightfoot Fl. Scot. p. 262.*

FRUTEX 6 pedalis et ultra, aculeatus, scandens, serpens.

A SHRUB six feet or more in height, prickly, climbing or creeping.

CAULIS teres, viridis, seu purpureus, ramosus, aculeatus, aculeis validis, recurvis, juniores rubicundi, seniores cinerei.

STALK round, green, or purple, branched and prickly, prickles strong, crooked back, the young ones bright red, the old ones ash-coloured.

FOLIA alterna, pinnata, plerumque septena, inodora, foliola sessilibus, ovalis, acutis, serratis, superne nitidis, inferne pallidioribus, infrioribus sensim minoribus, nervo medio foliata aculeato.

LEAVES alternate, pinnated, consisting for the most part of seven folioli, which are scentless, ovate, pointed, serrated, the upper side shining, the lower side paler, the lowermost ones gradually smaller, the mid-rib prickly underneath.

STIPULÆ denticulatae, denticulis apice rubris, capitatis.

STIPULÆ finely toothed, the teeth tipped with red, and terminated by a globule.

FLORES terminales, bini seu terni, etiam seni, pedunculati, pedunculis teretibus, nudis.

FLOWERS terminal, growing two or three, even sometimes six together, standing on footstalks, which are round and naked.

CALYX: calycis foliolis lanceolata, longe caudata, duo simplicia, duo utrinque pinnata, pinnis lanceolatis, acutis, unum ab altero tantum latere pinnatum, *fig.* 1.

CALYX: the foliola lanceolate, and long-tailed, two of them simple, two pinnated on each side, the pinnæ broadish and pointed, and one pinnated only on one side, *fig.* 1.

COROLLA: PETALA quinque, obcordata, remotiuscula, carnea, ad basin pallidiora.

COROLLA: five PETALS inversely cordate, a little remote from each other, pale red, fainter towards the base.

STAMINA: FILAMENTA plurima, lutea, setacea. ANTHERÆ incumbentes, ovatæ, *fig.* 2.

STAMINA: FILAMENTS numerous, yellow, tapering. ANTHERÆ incumbent, and ovate, *fig.* 2.

PISTILLUM: GERMINA plurima, intra tubum calycis, *fig.* 3. oblonga, lanata. STYLI: filiformes. STIGMATA plurima, sessile conniventia in capitulum, *fig.* 3.

PISTILLUM: GERMINA numerous, within the tube of the calyx, *fig.* 3. oblong and woolly. STYLES filiform. STIGMATA numerous, closely uniting and forming a little head, *fig.* 3.

PERICARPIUM Bacca ovalis, nitida, coccinea, unilocularis.

SEED-VESSEL: an oval, shining, scarlet BERRY of one cavity.

SEMINA plurima, luteſcentia, ſubovata, lanata, apice barbata.

SEEDS numerous, yellowish, somewhat ovate, woolly, bearded at top.

We remember somewhere to have seen an attempt to verify the Genera Plantarum: should such a plan ever be seriously agitated, we might recommend the following lines, written perhaps before any true notion was entertained of genus or species, as expressive of the Rose:

> " Quinque genus fratres, sub eodem tempore nati,
> " Bini barbati, bini fine arte creati,
> " Quintus habet barbam, sed tantum dimidiatam."

On examination it will appear, that this description, however quaint, accords exactly with the calyx in most, it not all, the species of this genus.

In some parts of Europe, particularly Austria and Carniola, the Roses are much more numerous than with us; and appear to create difficulties in determining the species to which we are happily strangers. Scopoli thus exclaims: " Fungos et Rosas quisque novit, species vero genuinas veriusque genuras ne Botanici quidem confirmant." The present species, without some little attention, may however be mistaken for the *alba*, especially when its flowers are whiter than ordinary.

The Dog Rose is well known to produce the Hep, a fruit agreeable enough when ripe and mellowed by the frost. Of this a conserve is made, and kept in the shops, where it is more used as a vehicle for other medicines than for any virtue of its own.

A very singular mossy protuberance is often found on various parts of this Rose, which is occasioned by an insect, the *Cynips Rosæ* of LINNÆUS. Formerly this substance, under the name *Bedeguar*, was used medicinally; but is now with much propriety rejected.

Its lively blossoms decorate our hedges in the month of July. The fruit is late before it ripens. In the winter it is much sought after by many birds, especially the Pheasant.

The water distilled from the wild Rose is said to be infinitely more fragrant than the common Rose water. HALLER says of it, " Fragrantia ejus olei omnia alia odoramenta superat, ut inter regia daus sit."

The strong thorns with which this shrub is furnished make it valuable either for forming hedges of itself, or for planting with others of stronger growth. The best way of raising plants for this purpose will be from seeds.

Rosa canina.

Tormentilla officinalis.

TORMENTILLA OFFICINALIS. TORMENTIL.

TORMENTILLA *Lin. Gen. Pl.* ICOSANDRIA POLYGYNIA.

             Cal. 8-fidus. *Petala* 4. *Sem.* fubrotunda, nuda, receptaculo
             parvo exfucro affixa.

    Raii Syn. Gen. 15. HERBA SEMINE NUDO POLYSPERMA.

TORMENTILLA *officinalis.*

TORMENTILLA *erecta caule erectiufculo, foliis feffilibus. Lin. Syft. Vegetab. p.399. Sp. Pl. p. 716.*
    Fl. Suec. n. 459.

FRAGRARIA tetrapetala, foliis caulinis feffilibus, quinatis. *Haller. hift. n. 1117.*

POTENTILLA *Tormentilla erecta. Scopoli Fl. Carn. n. 620.*

TORMENTILLA fylveftris. *Bauh. Pin.* 326.

TORMENTILLA *Ger. emac.* 992. vulgaris *Parkinf.* 394.
    Raii Syn. p. 257. Tormentil, Septfoil. *Hudson Fl. Angl. ed.* 2. *p.* 225.
    Lightfoot Fl. Scot. p. 272.

RADIX craffa, tuberofa, variæ magnitudinis et formæ, extus fufca, intus rubicunda.

CAULES plures ex una radice, fpithamæus et ultro, procumbentes, teretes, filiformes, pilofi, inferne fimplices, et fæpe nudi, fuperne ramofi.

FOLIA alterna, feffilia, utrapicauli-perfoliata, trifida, utrinque parce pubefcentia, fupra faturate viridia, lacinia obverfe lanceolatis, obtufis, fuperne laturibus, incifis, patentibus, tribus exterioribus duplo longioribus.

PEDUNCULI axillares, filiformes, elongati, unifiori, nudi, pilofi.

FLORES primo cernui, poftea erecti.

CALYX: PERIANTHIUM monophyllum, **octoparti**tum, pubefcens, laciniis ovatis, **acutis**, patentibus, alternis minoribus, *fig.* 1.

COROLLA: PETALA quatuor, lutea, obcordata, plana, patentia, unguibus calyci inferris, *fig.* 2.

STAMINA: FILAMENTA fedecim circiter, calyci inferta, corolla breviora: ANTHERÆ fimplices, luteæ, *fig.* 3.

PISTILLUM: GERMINA octo circiter, glabra, fubrotunda, in capitulum convenientia, *fig.* 4. STYLI filiformes, longitudine ftaminum, lateri germinis inferti; STIGMATA obtufa, *fig.* 5. auct.

RECEPTACULUM villofum.

SEMINA tot quot germina, oblongiufcula, obtufa, glabra, nuda, lutefcentia, *fig.* 6.

ROOT thick, and tuberous, various both in fize and fhape, externally brown, internally red.

STALKS feveral from one root, a fpan or more in length, procumbent, round, filiform, hairy, below fimple and often naked, above branched.

LEAVES alternate, feffile, nearly perfoliate, on each fide flightly pubefcent, above of a deep green colour, divided into many fegments, the fegments feverely lanceolate, obtufe, broadifh above, ferrated on the edges, and fpreading, the three outermoft twice as long as the others.

FLOWER-STALKS axillary, filiform, long, fupporting one flower, naked, and hairy.

FLOWERS at firft drooping, afterwards upright.

CALYX: PERIANTHIUM of one leaf, deeply divided into eight fegments, downy, the fegments ovate, pointed, alternately leaft, *fig.* 1.

COROLLA: four PETALS, of a yellow colour, inverfely heart-fhaped, flat, fpreading, inferted by the claws into the calyx, *fig.* 2.

STAMINA: about fixteen FILAMENTS, inferted into the calyx, fhorter than the corolla: ANTHERÆ fimple and yellow, *fig.* 3.

PISTILLUM: GERMINA about eight, fmooth, roundifh, forming a little head, *fig.* 4. STYLES filiform, the length of the ftamina, inferted into the fide of the germina; STIGMA blunt, *fig.* 5. magnified.

RECEPTACLE villous.

SEEDS as numerous as the germina, rather oblong, obtufe, fmooth, naked, and yellowifh, *fig.* 6.

Tormentil is a plant of confiderable importance in rural œconomy and medicine.

The roots are ufed in moft of the Weftern Ifles, and in the Orkneys, for tanning of leather; in which intention they are proved, by fome late experiments, to be fuperior even to the oak-bark. They are firft of all boiled in water, and the leather afterwards fteeped in the cold liquor. In the iflands of Tirey and Col the inhabitants have deftroyed fo much ground by digging them up, that they have lately been prohibited the ufe of them. *Lightfoot Fl. Scot. p.* 272.

Confidered medicinally, Tormentil root is a ftrong and almoft flavourlefs aftringent, and gives out its aftringency both to water and rectified fpirit, moft perfectly to the latter: the watery decoction, of a tranfparent brownifh-red colour whilft hot, becomes turbid in cooling like that of the Peruvian bark, and depofits a portion of refinous matter; the fpirituous tincture, of a brighter reddifh colour, retains its pellucidity. The extracts obtained by infpiffation, are intenfely ftyptic, the fpirituous moft fo. It is generally given in decoction: an ounce and a half of the powdered root may be boiled in three pints of water to a quart, adding, towards the end of the boiling, a drachm of cinnamon: of the ftrained liquor, fweetened with an ounce of any agreeable fyrup, two ounces or more may be taken four or five times a day.

We are by no means fond of changing the Linnæan names, but on the prefent occafion we are, in fome degree, compelled to it, from the great inconvenience we have experienced in calling a plant erecta, which with us is always procumbent, unlefs drawn up by furrounding herbage, or by growing in woods, where it more rarely occurs.

Its moft ufual place of growth is on heaths, moors, and mountainous paftures, where it is extremely common, and flowers from June to September.

LINNÆUS appears to have been induced to call this plant erecta, by way of contraft to the Tormentilla reptans, which he enumerates as a fpecies: fuch a plant is certainly figured and defcribed by feveral Englifh Botanifts, but we never yet faw any fpecies of Tormentil with a creeping ftalk: we have obferved the common Tormentil vary much in fize, in the length of its branches, and in the number and fize of its petals, we have noticed the leaves fometimes to have foot-ftalks, and we have for feveral years cultivated a large variety of this plant, which from one root has extended its ftalks nearly a yard every way, and though they have lain clofe to the ground, on a moift foil, we never could perceive the leaft tendency in them to throw out roots at the joints; hence we are induced to conclude, that no other than one fpecies of Tormentil exifts.

As the Tormentil varies with five petals, fo the Potentilla reptans has fometimes only four, and, perhaps, a ftarved fpecimen of the latter, originally gave rife to the Tormentilla reptans.

This occafional variation in the number of the petals, &c. at once deftroys the generic character of the Tormentil; for, add one fifth part more of the fructification to thofe which already exift in the Tormentilla, and you make a Potentilla of it; or, take one fifth-part of the fructification from a Potentilla, and it becomes a Tormentilla: they ought furely then to form but one genus: SCOPOLI makes them, facetioufly remarking, *Moneculam Hominem ab humano genere quis feparabit*: HALLER joins the Potentilla, Tormentilla, Fragraria, and Sibbaldia, in one family.

CISTUS HELIANTHEMUM. DWARF CISTUS.

CISTUS *Lin. Gen. Pl.* POLYANDRIA MONOGYNIA.

Cor. 5-petala. Cal. 5-phyllus; foliolis duobus minoribus. *Capfula.*

Raii Syn. Gen. 24. HEABE PENTAPETALÆ VASCULIFERÆ.

CISTUS *Helianthemum* fuffruticofus procumbens, ftipulis lanceolatis, foliis oblongis revolutis fubpilofis. *Lin. Syft. Vegetab.* Sp. Pl. 744. *Fl. Suec.* n. 472.

CISTUS foliis conjugatis, ellipticis, hirfutis, integerrimis, petiolis unifloris, fubhirfutis. *Hall. Hift.* 1033.

CISTUS *Helianthemum. Scopoli Fl. Carn.* n. 649.

CHAMÆ CISTUS vulgaris flore luteo. *Bauh.* p. 465.

HELIANTHEMUM Anglicum luteum. *Ger. em.* 1282.

HELIANTHEMUM vulgare. *Parkins.* 656. *Raii Syn.* p. 341. Dwarf Ciftus, or little Sun-Flower. *Hudfon Fl. Angl. ed.* 2. p. 233. *Lightfoot Fl. Scot.* p. 281. *Oeder Fl. Dan.* 101.

RADIX perennis, fubligofa, fufca.	ROOT perennial, fomewhat woody and brown.
CAULES plurimi, fuffruticofi, procumbentes, teretes, inferne glabri, fuperne hirfutuli, faepius rubicundi.	STALKS numerous, fomewhat fhrubby, procumbent, round, below fmooth, above flightly hairy, moft commonly reddifh.
FOLIA oppofita, breviffime petiolata, oblongo-ovata, acutiufcula, marginibus fubrevolutis, fuperne faturate viridia, fcabriufcula, fubpilofa, pilis furcatis, inferne fubtomentofa, *fig.* 1.	LEAVES oppofite, ftanding on very fhort foot ftalks, of an oblong ovate fhape, fomewhat pointed, the edges flightly rolled back, on the upper fide of a deep green colour, roughifh, and fomewhat hairy, the hairs forked, on the under fide a little downy. *fig.* 1.
STIPULÆ quaternae, lanceolatae, pilofae.	STIPULÆ growing four together, lanceolate, and hairy.
CALYX: PERIANTHIUM pentaphyllum, perfiftens, foliolis tribus fuperioribus ovatis, obtufiufculis, membranaceis, fubdiaphanis, aequalibus, concavis, trinerviis, nervis coloratis, hirfutulis, duobus inferioribus minutis, lateralibus hirfutis, *fig.* 2, 3.	CALYX: a PERIANTHIUM of five leaves and permanent, the three uppermoft ones ovate, bluntifh, membranaceous, fomewhat tranfparent, equal, concave, three-ribbed, the ribs coloured and hairy, the two lowermoft very fmall, lateral, and hairy, *fig.* 2, 3.
COROLLA: PETALA quinque obcordata, flava, margine exteriore crenulata, *fig.* 4.	COROLLA: five PETALS inverfely heart-fhaped, of a yellow colour, the outer edge flightly notched, *fig.* 4.
STAMINA: FILAMENTA numerofa, capillaria, flava, receptaculo fupra calycem inferta. ANTHERAE fubrotundae, parvae, flavae, *fig.* 5.	STAMINA: FILAMENTS numerous, capillary, yellow, inferted into the receptacle above the calyx. ANTHERAE roundifh, fmall, and yellow, *fig.* 5.
PISTILLUM: GERMEN fubrotundum. STYLUS longitudine ftaminum, fuperne craffior, inferne faepius curvatus. STIGMA capitatum, planum, *fig.* 6.	PISTILLUM: GERMEN roundifh. STYLE the length of the ftamina, thicker in its upper part, and crooked below. STIGMA forming a little flat head, *fig.* 6.
PERICARPIUM: CAPSULA fubrotunda, calyce tecta, uniloccularis, trivalvis, *fig.* 7.	SEED-VESSEL: a roundifh CAPSULE, covered with the calyx, of one cavity and three valves, *fig.* 7.
SEMINA plurima, majufcula, ovato-acuta, rufa, *fig.* 8.	SEEDS numerous, rather large, ovate, pointed, and of a reddifh brown colour, *fig.* 8.

Moft of the plants of the Ciftus tribe are highly efteemed for their beauty, and generally cultivated in the gardens of the curious. Though our prefent fpecies cannot vie with many of thofe which are the produce of warmer climates, yet it is one of the moft ornamental of our native plants, and admirably well calculated to decorate a rock or dry bank, efpecially if its feveral varieties with white, rofe, and lemon-coloured flowers be intermixed. The particular merit of this plant is, that it is hardy, eafily propagated, either by feeds or cuttings, and continues for the greateft part of the fummer to put forth daily a multitude of new bloffoms.

Mr. LAWSON is faid by Mr. RAY to have found it producing white flowers. I have myfelf obferved a wild variety with pale yellow bloffoms. A variety with double flowers is mentioned by HALLER, which, if it could be procured, would be a valuable acquifition to our gardens. LINNÆUS has remarked, that the petals fometimes have an orange-coloured fpot at their bafe; and the leaves have been obferved to vary much in breadth.

In chalky foils the *Ciftus Helianthemum* is extremely common; but as that does not abound in the neighbourhood of London, it is confequently fcarce with us.

On a clofe examination of the hairs on the leaves we difcovered them to be forked; a character which may, perhaps, contribute to diftinguifh it from the *polifolia*, to which it forms very nearly related.

It flowers from *June* to *Auguft*.

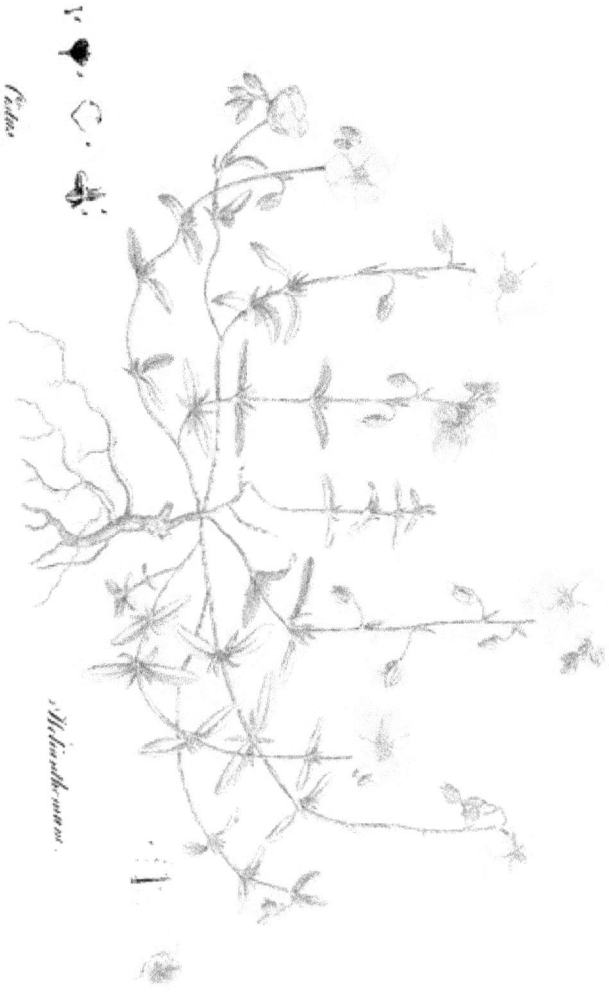

Papaver dubium

PAPAVER *Lin. Gen.* IV. POLYANDRIA MONOGYNIA.

Cor. 4-petala. *Cal.* 2-phyllus. *Capfula* 1-locularis, fub ftigmate perfiftente poris dehifcens.

Raii Syn Gen. 12. HERBÆ VASCULIFERÆ FLORE TETRAPETALO ANOMALÆ.

PAPAVER *dubium capfulis oblongis glabris, caule multifloro fetis afperfis, foliis pinnatifidis incifis.* *Lin. Syft. Vegetab.* p. 407. *Sp. Pl.* 726. *Fl. Suec. n.* 467.

PAPAVER *foliis hifpidis, pinnatis, pinnis lobatis, fructu ovato lævi.* *Haller. Hift. n.* 1063.

PAPAVER *erraticum capite longiffimo glabro.* *Tourn. Inft.* 238.

PAPAVER *laciniato folio, capitulo longiore glabro, feu Argemone capitulo longiore glabro.* *Mor. H. R. Bl. H. Ox. II.* 279. *S. III. t.* 14. *fig.* 11. *Raii Syn.* p. 309. Smooth-headed Baftard-Poppy. *Hudfon. Fl. Angl.* p. 231. *Lightfoot Fl. Scot.* p. 280.

This plant, in its general appearance, is fo very fimilar to the *Papaver Rhœas,* as often to be overlooked and miftaken for that fpecies. Were the flowers white, as Jacquin informs us they conftantly are in Auftria, the two plants would be much more obvioufly diftinguifhed; but, fortunately, it has a few characters which always point it out to the attentive obferver. Thefe are principally drawn from the Capfules and Flower-ftalks; the Capfules of the *Rhœas* are broad and fhort, fomewhat refembling one-half of an egg cut tranfverfely; thofe of the *dubium* are long and flender. Such is the general appearance of the two Capfules, which, however, are fubject to confiderable variation. In the *Rhœas,* the hairs on the Flower-ftalk are ftrong, rigid, and fpread horizontally; in the *dubium* they are finer, and preffed upward clofe to the ftalk *. On the young Flower-ftalks they affume a fhining, filvery-white appearance, which looks very beautiful. Below the Flower-ftalks, on the other parts of the plant, the hairs fpread out. In this laft character we do not recollect to have ever been deceived. Befides thefe, which are the principal differences, the ftalks and leaves of the *dubium* are much paler, the flowers are alfo much fmaller, and lefs intenfely red.

Culture produces no alteration in the conftancy of its characters.

In Battersea Fields, where the foil is light, the *dubium* is nearly as common, and as much of a weed, as the *Rhœas;* nor is it unfrequent on walls, in the environs of the Metropolis; according to Mr. LIGHTFOOT, it is the moft common fpecies in North Britain.

In a corn field, betwixt Croydon and Shirley Common, we once noticed feveral fpecimens of this poppy with very large Capfules, which, if we miftake not, were difeafed.

It flowers in June.

* Jacquin's figure reprefents the hairs of the Flower-ftalks reverfed, and the leaves too finely divided.

PAPAVER *Lin. Gen. Pl.* Polyandria Monogynia.

> *Cor.* 4 petala. *Cal.* 2 phyllus. *Capfula* 1-locularis, fub ftigmate perfiftente poris dehifcens.

Raii Syn. Gen. 22. Herbæ vasculiferæ, flore tetrapetalo anomalæ.

PAPAVER *Argemone* capfulis clavatis hifpidis, caule foliofo multiflore. *Lin. Syft. Vegetab.* p. 407. *Spec. Pl.* 725. *Fl. Suec. n.* 466.

PAPAVER foliis hifpidis, pinnatis, pinnis lobatis, capitulis ellipticis, hifpidis. *Haller Hift. n.* 1063.

PAPAVER *Argemone.* *Scopoli Fl. Carn. n.* 636.

ARGEMONE capitulo longiore. *C. Bauh. Pin.* 172. *Ger. emac.* 273. *Park.* 370.

PAPAVER laciniato folio, capitulo hifpido longiore. *Raii Syn.* p. 308. Long rough-headed baftard Poppy. *Hudfon Fl. Angl. ed. 2. p.* 230. *Lightfoot Fl. Scot. p.* 279.

RADIX annua, fimplex, fibrofa.

CAULIS: ubi læte crefcit exoluo proffert plures, pediales, et ultra, foliofos, adfcendentes, hirfutos, inter fegetes vero caule folitario erecto fæpius gaudet.

FOLIA radicalia plurima, longe petiolata, pinnata, pinnis incifo-dentatis, dentibus mucronatis, caulina tripartita, pinnatifida, omnibus pilofis, fuperne faturate viridibus, nitidis, inferne pallidioribus.

PEDUNCULI pilofi, pilis adpreffis.

CALYX: Perianthium diphyllum, feu triphyllum, deciduum, papillofo-hifpidum.

COROLLA: Petala quatuor, miniata, fuberecta, remotiufcula, obverfe-ovata, apice crenulata, bafi nigricantis, maxime caduca, *fig.* 1.

STAMINA: Filamenta viginti circiter, purpurea, plana, apice dilatata, nitida. Antheræ breviffime pedicellatæ, biloculares. Pollen cærulefcens, *fig.* 2. auct. *fig.* 3.

PISTILLUM: Germen longitudine filamentorum, clavatum, fubangulatum, hifpidum, pilis canis, adpreffis. Stigmatis radii 3 ad 5 villofi, cærulefcentes, *fig.* 4.

PERICARPIUM: Capfula oblonga, clavata, fubangulofa, hifpida, inferne medio nuda, purpurafcens, *fig.* 5.

SEMINA plurima, minuta, nigricantia, *fig.* 6, 7.

ROOT annual, fimple, and fibrous.

STALK: where the plant grows luxuriantly, it puts forth feveral leafy, hairy ftalks, a foot or more in height, and bending upwards, but among corn it is moft commonly found with a fingle upright ftem.

LEAVES next the root numerous, ftanding on long foot-ftalks, pinnated, the pinnæ deeply indented, the teeth terminating in a fhort point, thofe of the ftalk deeply divided into three fegments which are pinnatifid, all the leaves are hairy, on the upper fide of a deep green colour, and fhining, on the underfide paler.

FLOWER-STALKS hairy, hairs preffed clofe to the ftalk.

CALYX: a Perianthium compofed of two or three leaves, deciduous, hifpid, the hairs iffuing from fmall papillæ or prominent points.

COROLLA: four Petals, of a fcarlet colour, nearly upright, a little diftant from each other, inverfely ovate, finely notched at top, and blackifh at the bafe, *fig.* 6.

STAMINA: about twenty Filaments, of a purple colour, flat, dilated at top, and fhining. Antheræ ftanding each on a very fhort foot-ftalk, having two cavities. Pollen blueifh, *fig.* 2. one of the ftamina magnified, *fig.* 3.

PISTILLUM: Germen the length of the filaments, thickeft at top, fomewhat angular, hifpid, the hairs grey and preffed to it. Stigma compofed of 3 to 5 villous rays, of a blueifh colour, *fig.* 4.

SEED-VESSEL: an oblong, club-fhaped Capfule, fomewhat angular, hifpid, below for the moft part naked, of a purplifh colour, *fig.* 5.

SEEDS numerous, minute, and blackifh, *fig.* 6, 7.

This fpecies of Poppy is diftinguifhed by a variety of particulars befides its long prickly heads, which, though not abfolutely neceffary to difcriminate the fpecies, are well worthy of our attention. The divifions of the leaves are finer than in any of the other poppies. The petals in general grow more upright: and, inftead of having the edges falling over each other, are ufually a little diftant. The ftamina are very remarkable, having the filaments uncommonly dilated towards the top, not at the bafe, as Haller afferts; and the Antheræ ftand on a very flender foot-ftalk placed on the top of each filament.

Like moft of the other poppies it ufually grows in corn fields, and is not very unfrequent in the neighbourhood of London. About the beginning of June it bloffoms in Batterfea Fields; but it is often overlooked from the extreme caducity of its petals, which rarely continue expanded more than fix hours.

Papaver Argemone

ORIGANUM VULGARE. WILD MARJORAM.

ORIGANUM. *Lin. Gen. Pl.* DIDYNAMIA GYMNOSPERMIA.

Strobilus tetragonus, *spicatus, calyces* colligens. *fig. 6.*

Rail Synop. Gen. 14. SUFFRUTICES ET HERBÆ VERTICILLATÆ.

ORIGANUM vulgare *spicis* subrotundis paniculatis conglomeratis, bracteis calyce longioribus ovatis. *Lin. Syst. Vegetab.* p. 157. *Spec. Pl.* p. 824. *Fl. Suec.* n. 534.

ORIGANUM foliis ovatis, umbellis comosis, flaminibus exsertis. *Haller hist.* n. 233.

ORIGANUM vulgare. *Scopoli Fl. Carn.* n. 740.

ORIGANUM sylvestre. *Bauh. pin.* 223.

ORIGANUM anglicum. *Ger. emac.* 665.

MAJORANA sylvestris. *Park.* 12.

ORIGANUM vulgare spontaneum. *Bauh. hist. III.* 236.

Raii Syn. 236. Wild Marjoram. *Hudson Fl. Angl. ed. 2. p. 262. Lightfoot Fl. Scot. p. 317.*

RADIX perennis, repens, horizontalis, fusca, plurimis fibris capillata.

CAULIS pedalis, ad sesquipedalem, erectus, tetragonus, purpurascens, pubescens, ramosus.

RAMI oppositi, erecti, caule teneriores, in cæteris conformes.

FOLIA ad genicula, oppositis, petiolata, ovata, acuta, minutius et rarius denticulata, supra glabriuscula, subtus pubescentia, utrinque punctata, margine minutim ciliata, perennia.

PETIOLI pubescentes.

AXILLÆ foliorum in planta culta foliolis onustæ.

FLORES paniculati, panicula e spicis pluribus, subrotundis, conglomeratis composita.

BRACTEÆ ovato-lanceolatæ, sessiles, concavæ, integræ, corolla integris coloratæ, ad lentem pubescentes, floribus subjectæ singulæ, fig. 1.

CALYX: PERIANTHIUM monophyllum, tubulatum, striatum, subpubescens, pedicellatum, longitudine fere bracteæ, ore barbato, quinquefido, laciniis acutis, erectis, æqualibus, purpureis. fig. 2.

COROLLA infundibuliformis, purpurea, tubus villosus, sensim sursus ampliatus, calyce longior, limbus bilabiatus, labium superius evolutum, bifidum, obtusum, inferius trifidum, patens, obtusum, fig. 3.

STAMINA: FILAMENTA quatuor, purpurea, corolla paulo longiora, duobus inferioribus paulo longioribus; ANTHERÆ didymæ, saturatius coloratæ, fig. 4.

PISTILLUM: GERMEN quadripartitum; STYLUS filiformis, corolla longior. STIGMA bifidum, acutum, revolutum, fig. 5.

SEMINA quatuor, ovata, in sinu calycis conniventia.

ROOT perennial, creeping, horizontal, brown, tufted with numerous fibres.

STALK a foot or a foot and a half high, upright, four cornered, purplish, downy, and branched.

BRANCHES opposite, upright, more tender than the stalk, in other respects similar.

LEAVES placed at the joints, opposite, standing on foot-stalks, ovate, pointed, finely and rarely toothed, above nearly smooth, beneath downy, dotted on both sides, the edge finely fringed, spreading.

LEAF-STALKS downy.

ALÆ of the leaves, in the cultivated plant, bearing numerous small leaves.

FLOWERS forming a panicle, composed of numerous roundish spikes, growing in clusters.

FLORAL-LEAVES ovato-lanceolate, sessile, concave, entire, more deeply coloured than the corolla, appearing downy when magnified, placed one under each flower, fig. 1.

CALYX: A PERIANTHIUM of one leaf, tubular, striated, slightly downy, standing on a short foot-stalk, and almost the length of the floral-leaf, the mouth bearded, divided into five, pointed, upright, equal, purple segments, fig. 2.

COROLLA funnel-shaped, purple, the tube villous, gradually enlarged upwards, longer than the calyx, the limb composed of two lips, the upper lip upright, bifid and obtuse, the lower lip trifid, spreading and obtuse, fig. 3.

STAMINA: four purple FILAMENTS, a little longer than the corolla, the two lowermost somewhat the longest; ANTHERÆ double, and more deeply coloured, fig. 4.

PISTILLUM: GERMEN divided into four parts. STYLE filiform, longer than the corolla; STIGMA bifid, pointed, and turned back, fig. 5.

SEEDS four, ovate, in the bottom of the calyx, which closes over them.

This aromatic and ornamental plant, grows wild on dry chalky hills, and gravelly ground, in most parts of Great Britain, though sparingly in the vicinity of London.

It flowers in July and August.

The leaves and flowery tops of Origanum have an agreeable aromatic smell, and a pungent taste, warmer than that of the Garden Marjoram, and much resembling Thyme, with which they appear to agree in medicinal virtue. Infusions of them are sometimes drank as tea, in weakness of the stomach, disorders of the breast, for promoting perspiration, and the fluid secretions in general; they are sometimes used also in nervous and antirheumatic baths; and the powder of the dried herb is an errhine. Distilled with water, they yield a moderate quantity of a very acrid and penetrating essential oil, smelling strongly of the Origanum, but less agreeable than the herb itself: this oil is applied on a little cotton for easing the pains of carious teeth; and sometimes diluted and rubbed on the nostrils, or snuffed up the nose, for astonishing and evacuating mucous humours. *Lewis M. Med.* p. 469.

It dyes linen cloth of a reddish brown colour; for this purpose the linen is first macerated in alum water and dried; it is then soaked for two days in a decoction of the bark of the crab-tree; it is wrung out of this, boiled in a ley of ashes, and then suffered to boil in the decoction. *Haller hist. Helv.* p. 193.

According to LINNÆUS, it dyes woollen cloth also of a purple colour; is sometimes used as a succedaneum for tea, and added to beer to make it more quickly intoxicate, as likewise to prevent it from too quickly turning sour.

Origanum vulgare.

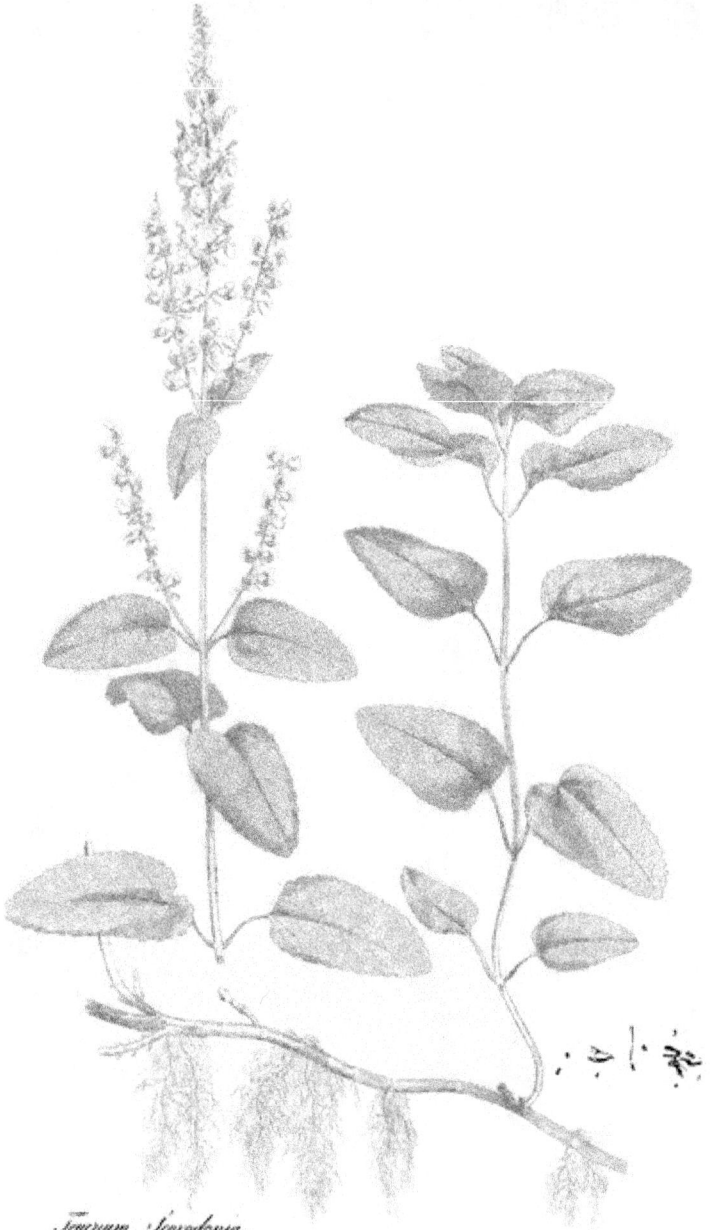

Teucrium Scorodonia.

Teucrium Scorodonia. Sage-leaved Germander, or Wood Sage.

TEUCRIUM *Lin. Gen. Pl.* Didynamia Gymnospermia.
Corolla labium superius (nullum) ultra basin bipartitum, divaricatum ubi staminia.

Raii Syn. Gen. 14. Superfloribus et Herba verticillatis.

TEUCRIUM *Scorodonia* foliis cordatis serratis petiolatis, racemis lateralibus secundis, caule erecto. *Lin. Syst. Vegetab.* p. 440. *Sp. Pl.* 789.

CHAMÆDRYS folio exotica prostrato, flore longitudine modo hexagonali. *Haller. Hist.* n. 287.

TEUCRIUM *Scordoum. Scopoli Fl. Carn.* n. 731.

SCORDIUM alterum sive folio agresti. *Bauh. Pin.* 247.

SCORODONIA sive folio agresti. *Ger. em.* 662.

SCORODONIA Scordium flotum odoratum et folio agresti. *Park. 111. Raii Syn.* 243. *Haller. Fl. Angl.* p. 228. *Lightfoot Fl. Scot.* p. 303. *Fl. Dan. t.* 485.

RADIX perennis, lignosa, subrepens.	ROOT perennial, woody, and somewhat creeping.
CAULES plures, sesquipedales, bipedales et ultra, subterecti, tetragoni, duri, purpurei, hirsuti.	STALKS several, a foot and a half, two feet high, and more, nearly upright, four-cornered, hard, purple, and hairy.
FOLIA opposita, petiolata, cordato-oblonga, plerumque obtusa, saepe vero rotundata, utrinque subvenosa, utrinque utrinque, obtusiter et conjunctiter serrata.	LEAVES opposite, standing on foot-stalks, of an oblong heart form, generally obtuse, but often a little pointed, veiny like sage, a little hairy on each side, obtusely and conjointly serrated.
PETIOLI hirsuti.	LEAF-STALKS hairy.
FLORES stramineae, racemosi, secundi, racemis oppositis, longis, nudis, terminali duplo longiore.	FLOWERS straw-coloured, growing all one way, on long, opposite, naked racemi, the terminal one of which is almost twice as long as the rest.
BRACTÆA ovato-acuminata, singulo flori subjecta.	FLORAL-LEAF ovate, pointed, and placed under each flower.
CALYX: Perianthium monophyllum, tubulosum, inferne basi gibbosum, labio superiore erecto, integro, vel obsolete trilobo; inferiore quadridentato, dentibus triangularibus, *fig.* 1.	CALYX: a Perianthium of one leaf, tubular, on the under side gibbous at the base, the upper lip upright, entire, or faintly three-lobed; the lower lip furnished with four teeth, which are nearly equal, *fig.* 1.
COROLLA monopetala, ringens: Tubus cylindraceus, brevis: Labium superius ultra basin profunde bipartitum, distantibus ad latera lateralia; Labium inferius patens, trifidum, laciniis lateralibus figura labii superioris, media maxima, subrotunda, *fig.* 2.	COROLLA monopetalous and ringent: Tube cylindrical and short: upper Lip deeply divided beyond the base, figments standing wide down each spreading, mild, lateral legments the same shape as the figments of the upper lip, the middle one very large and roundish, *fig.* 2.
STAMINA: Filamenta quatuor, quorum duo longiora, purpurea, prius, prima erecta, conniventia, postea reflexa, et disjuncta: Antheræ flavæ, *fig.* 3.	STAMINA: four Filaments, two of which are longer than the rest, purple and hairy, at first upright and closing together, afterwards turned back, and separated: Antheræ yellow, *fig.* 3.
PISTILLUM: Germen quadripartitum. Stylus filiformis. Stigmata duo, tenuia, *fig.* 4.	PISTILLUM: Germen quadripartite, Style filiform. Stigmata two, slender, *fig.* 4.
SEMINA quatuor, subrotunda, angulata, nitida, in fundis calycis, pilis transversis rigidis tecta, ibique detenta, ad debitam maturitatem, *fig.* 5.	SEEDS four, nearly round, blackish, shining, in the bottom of the calyx, closely covered with coats rigid hairs, and kept there till they have acquired a proper degree of ripeness, *fig.* 5.

The Wood-sage, or more properly Sage-leaved Germander, delights to grow in woods and hilly situations, among bushes, and under hedges, where the soil is dry and stony; and in such places it is not only common with us, but frequent in most parts of Great Britain.

It flowers in July, August, and September.

Its leaves much resemble those of Sage, from which circumstance, and not from any botanical or medical affinity, it receives its name.

As a medicinal plant, it has never been highly celebrated. Lewis omits it in his Materia Medica, but retains it in his Dispensatory: its smell, taste, and medical virtues, he says, it comes nearer to Scordium than Sage. Rutty relates a case of Vertigo, brought on by the odour which arose from frequently handling the herb in the distillation of it. He mentions it as the smell of the Hop, in lieu of which, he says, it may be substituted in making beer; and that, when boiled in the wort, the beer sooner becomes clear than when hops are made use of. Its virtues, in this respect, are highly extolled by the Rev. P. Laurents of Bury*. We have only to wish, that experiment may justify the encomiums of our learned and benevolent friend.

"Seeing so much fine ground under costly hops, which, it must be owned, had very large and verdant leaves, I could not but repine at the expence of soil, poles, dung, and labour, bestowed on this plant, especially when there is great reason to suppose, that the *Teucrium Scorodonia* would better answer the purpose. Of this plant I can so far say, that in smell and taste it resembles Hops. The name by which it goes in some authors is *Ambrosia*, a name announcing something immortal and divine; and to this day, *ambrosia* is the appellation by which it goes among the common people in the island of Jersey. Here, where Cyder, the common beverage, has failed, I have known the people melt each his barley at home, and, instead of Hops use to very good purpose, the *ambrosia* of their hedges.

"It is my ardent wish, I own, to see justice done to the neglected merits of this unvalued plant; but should indolence, prejudice, or private interest, obstruct the introduction of it into use, let me at last entreat brewers to honour it with their notice, in preference to any unpalatable and wholesome substitute they may have occasion to use in lieu of Hops."

* Vide Young through Flanders, &c. published in the fourth volume of the Views of Annals of Agriculture.

ANTIRRHINUM MINUS. THE LEAST TOAD-FLAX.

ANTIRRHINUM *Lin. Gen. Pl.* DIDYNAMIA ANGIOSPERMIA.

Cal. 5-phyllus. *Corolla basis deorsum prominens, nectarifera. Capsula* 2-locularis.

Raii Syn. Gen. 18. HERBÆ FRUCTU SICCO SINGULARI FLORE MONOPETALO.

ANTIRRHINUM *minus foliis plerisque alternis lanceolatis obtusis, caule ramosissimo diffuso. Lin. Syst. Vegetab.* p. 466. *Sp. Pl.* p. 852. *Fl. Suec.* p. 502.

ANTIRRHINUM *viscidum foliis inferioribus conjugatis ellipticis obtusis hirsutis, calcare dimidii floris longitudine. Haller. Hist.* n. 335.

ANTIRRHINUM *minus. Scopoli Fl. Carn.* n. 769.

ANTIRRHINUM *arvense minus. Bauh. pin.* 212.

ANTIRRHINUM *minimum repens. Ger. emac.* 549.

ANTIRRHINUM *sylvestre minimum. Parkins.* 1334.

LINARIA *Antirrhinum dicta. Raii Syn.* p. 283. The least Calf's Snout or Snap-dragon. *Hudson. Fl. Angl. ed.* 2. p. 272. *Order. Fl. Dan. t.* 532.

RADIX annua, simplex, fibrosa.

CAULIS erectus, spithamæus, seu dodrantalis, ad basin usque ramosus, teres, ramis inferioribus oppositis, superioribus alternis.

FOLIA ut ut tota planta villosa, subviscosa, inferiora opposita, patentia, subspatulata, superiora alterna, recurvata, lineari-lanceolata, obtusa.

FLORES parvi, solitarii, alterni, pedunculati, pedunculis erectis.

CALYX: PERIANTHIUM quinque-partitum, persistens, laciniis linearibus, subæqualibus, corolla brevioribus, *fig.* 1.

COROLLA monopetala, tubus superne purpureus, inferne maculis duabus parallelis, purpureis notatus, calcare brevissimum tubulatum purpurascens, labium superius bifidum, inferius albidum, inferius trifidum, album; palatum villosum, flavescens, *fig.* 2.

STAMINA: FILAMENTA quatuor, alba. ANTHERÆ nigricantes. POLLEN album.

PISTILLUM: GERMEN subovatum, viscidum, rufescens. STYLUS filiformis, superne purpureus. STIGMA simplex, album.

PERICARPIUM: CAPSULA ovata, apice dehiscens.

ROOT annual, simple, and fibrous.

STALK upright, from five to nine inches in height, branched down to the bottom, round, the lowermost branches opposite, the uppermost alternate.

LEAVES as well as the whole plant villous, and somewhat viscid, the lower ones opposite, spreading, somewhat spatula-shaped, the upper ones alternate, bent back, betwixt linear and lanceolate, the extremity obtuse.

FLOWERS small, solitary, alternate, standing on upright foot-stalks.

CALYX: a PERIANTHIUM deeply divided into five segments, which are linear, nearly equal, shorter than the corolla and permanent, *fig.* 1.

COROLLA monopetalous, the tube on the upper side purple, underneath marked with two parallel purple spots, spur very short and tapering, of a purplish colour, the upper lip bifid, on the underside whitish, the lower trifid and white, the palate villous and yellowish, *fig.* 2.

STAMINA: four white FILAMENTS. ANTHERÆ blackish. POLLEN white.

PISTILLUM: GERMEN somewhat ovate, viscid, and of a reddish brown colour. STYLE filiform, on the upper part purplish. STIGMA simple and white.

SEED-VESSEL, an ovate CAPSULE opening at top.

Botanists have distinguished this species by the names of *minus* and *minimum*, as being the most diminutive of the genus. It may also be considered as one of the least ornamental.

It is chiefly found in corn fields, especially where the soil is sandy. We have occasionally noticed it in Battersea Fields with the *Orvala*; but in many parts of Kent it grows much more plentifully.

We know of no use to which it is applicable; and it is too diminutive a plant to do much harm where it is most abundant.

Introduced into the garden, it comes up annually without any care, nor is it easily lost.

It branches and spreads according to the luxuriance of the soil, and frequently grows to a much greater size than our figure represents.

It flowers from June to August.

Asterolinum minus.

Euphrasia officinalis

EUPHRASIA OFFICINALIS. COMMON EYEBRIGHT.

EUPHRASIA *Lin. Gen. Pl.* DIDYNAMIA ANGIOSPERMIA.

Cal. 4-fidus, cylindricus. *Caps.* 2-locularis, ovato-oblonga. *Antheræ* inferiores altero lobo basi spinosæ.

Raii Syn. Gen. 18. HERBÆ FRUCTU SICCO SINGULARI FLORE MONOPETALO.

EUPHRASIA *officinalis* foliis ovatis lineatis argute dentatis. *Lin. Syst. Vegetab.* p. 460. *Sp. Pl.* p. 481. *Fl. Suec. n.* 513. *Haller hist.* 303.

EUPHRASIA *officinalis. Scopoli Fl. Carn. n.* 753.

EUPHRASIA officinarum. *Bauh. pin.* 233. *Ger. emac.* 663. *Parkins.* 1329. *Raii. Syn. p.* 284. Eyebright, *Hudson Fl. Angl. ed. 2. p.* 268. *Lightfoot Fl. Scot. p.* 303.

RADIX annua, fibrosa, albida.

CAULIS bipollicaris ad palmarem et ultra, erectus, teres, pubescens, purpureus, plerumque ramosus.

FOLIA opposita, ovata, obtusa, serrato-dentata, dentibus acuminatis, supra convexa, subtus concava, marginibus ciliatis, utrinque hirsutula, supra nitidula, inæsta, subtus venosa.

RACEMUS terminalis, foliaceus, erectus, floribus axillaribus, oppositis, sessilibus.

CALYX: PERIANTHIUM monophyllum, ovatum, angulatum, persistens, foliis paulo brevius, pubescens, quadrifidum, laciniis, lanceolatis, acuminatis, erectis, ciliatis, subæqualibus, *fig.* 1.

COROLLA monopetala, alba, ringens; *Tubus* cylindricus, albus, glaber, longitudine calycis, *fig.* 2. *Limbus* bilabiatus; *Labium* superius album, suberosum, concavum, pubescens, striis cæruleis utrinque 5, tetus petiorum, ebusdum, erectum, bilobum, lobis emarginatis, *fig.* 3; inferius 3-partitum paulo majus, trifidum, laciniis omnibus emarginatis, *fig.* 4. *Faux* undique striata, et picta lineis caruleiscentibus, antice vero colore luteo.

STAMINA: FILAMENTA quatuor, subulata, purpurascentia, tubo inserta, *fig.* 5. ANTHERÆ purpureæ, bilobæ, obtusæ, subtus barbatæ, conniventes, lobis spinula terminatis, duabus inferioribus longioribus, *fig.* 6, 7.

PISTILLUM: GERMEN ovatum, obtusum, barbatum, *fig.* 8. STYLUS filiformis, superne pubescens, *fig.* 9. STIGMA obtusum, integrum, *fig.* 10.

PERICARPIUM: CAPSULA ovato-oblonga, compressa, obtusa, mucronata, bilocularis, *fig.* 11.

SEMINA plurima, albida, striata, *fig.* 12.

ROOT annual, fibrous, and whitish.

STALK from two to four inches high, or more, upright, round, hoary, purple, for the most part branched.

LEAVES opposite, ovate, obtuse, serrated or indented, teeth pointed, above convex, beneath concave, finely edged with hairs, slightly hirsute on each side, above somewhat glossy, with lines impressed, underneath veiny.

RACEMUS terminal, leafy, upright, flowers in the ales of the leaves, opposite and sessile.

CALYX: a PERIANTHIUM of one leaf, ovate, angular, permanent, a little shorter than the leaves, pubescent, divided into four segments, which are lanceolate, long-pointed, upright, edged with hairs, and nearly equal, *fig.* 1.

COROLLA monopetalous, white, ringent; *Tube* cylindrical, white, smooth, the length of the calyx, *fig.* 2. *Limb* two-lipped; upper *Lip* white, somewhat ovate, hollow, downy, painted on the inside with three bluish streaks on each side, blunt, upright, bifid, the lobes emarginate, *fig.* 3. the lower lip somewhat larger than the upper, trifid, all the segments emarginate, *fig.* 4. Mouth streaked all round, and painted with bluish streaks, but anteriorly of a yellow colour.

STAMINA: four tapering, purplish FILAMENTS inserted into the tube of the corolla, *fig.* 5. ANTHERÆ purple, two-lobed, obtuse, bearded underneath, closing together, the lobes terminating in a spine, the two lowermost the longest, *fig.* 6, 7.

PISTILLUM: GERMEN ovate, obtuse, bearded, *fig.* 8. STYLE thread-form, downy, on the upper part, *fig.* 9. STIGMA blunt, and entire, *fig.* 10.

SEED VESSEL: an ovate, oblong, CAPSULE, flattened, obtuse, with a short point, of two cavities, *fig.* 11.

SEEDS several, whitish, and striated, *fig.* 12.

Eyebright is a very common plant on heaths, and pastures, especially where the soil is chalky; it varies much in size and in the branchedness of its stalk, as well as in the colour and size of its blossoms, and flowers from July to September.

Many writers on the Materia Medica, ascribe to this plant wonderful efficacy in disorders of the Eyes: ALSTON says, it has been long reckoned a specific opthalmic, and commended for dim, weak, and watery eyes, for inflamed and sore eyes, for cataracts, &c. yea, it is said to make old eyes become young again, and the blind to see. MILTON, who most probably from his own misfortune, had been induced to look into books of this sort, thus mentions it:

> "_____ but to nobler sights
> "Michael from Adam's eyes the film remov'd,
> "Which that false fruit that promis'd clearer sight
> "Had bred; then purg'd with *euphrasy* and rue
> "The visual nerve, for he had much to see."

On the other hand, there are not wanting those who condemn its use, especially in inflammatory complaints of the eyes; a friend of LOBEL's is said nearly to have lost his eyesight by the use of it. In such contrariety of sentiment, it will, perhaps, be most prudent not to lay too much stress on so doubtful a remedy.

RHINANTHUS CRISTA GALLI. YELLOW RATTLE.

RHINANTHUS *Lin. Gen. Pl.* DIDYNAMIA ANGIOSPERMIA.

Cal. 4-fidus, ventricosus. Capsula 2-locularis, obtusâ, compressâ.

Rai. Syn. Gen. 28. FLORES FRUCTU SICCO SINGULARI, FLORE MONOPETALO.

RHINANTHUS *Crista Galli* corollis labio superiore compresso breviore. *Lin. Syst. Vegetab. p.* 459.

ALECTOROLOPHUS calycibus glabris. *Haller. Hist.* 313.

MIMULUS *Crista Galli. Scopoli Fl. Carn. n.* 751.

PEDICULARIS *pratensis lutea vel Crista Galli. Baub. Pin.* 163.

CRISTA GALLI *foemina. I. B. III.* 436.

CRISTA GALLI *Ger. em.* 1071.

PEDICULARIS *seu Crista Galli lutea. Park.* 713. Yellow Rattle or Cocks-comb. *Rai Syn.* * 284. *Hudson. Fl. Angl. ed. 2. N.* 268. *Lightfoot Fl. Scot. p.* 322.

RADIX annua, simplex, albida, parum fibrosa.	ROOT annual, simple, whitish, furnished with few fibres.
CAULIS pedalis circiter, erectus, simplex, seu ramosus, quadrangulus, glaber, purpureo maculatus.	STALK about a foot high, upright, simple or branched, square, smooth, and spotted with purple.
FOLIA opposita, remotiuscula, sessilia, cordato-lanceolata, obtusiuscula, venosa, brevia, subtus tuberculis albidis pulchre reticulatis, serrata, serraturis imis glandulis et subincrassatis.	LEAVES opposite, rather remote from each other, sessile, lanceolate with a heart-shaped base, blueish, veiny, smooth, underneath beautifully reticulated with white tubercles, serrate, the notches thick on the edge, and somewhat rolled back.
BRACTEÆ oppositæ, magnæ, foliis similes ad basi latiores, et profundius incisæ, serraturis acuminatis.	FLORAL-LEAVES opposite, large like the leaves, but broader at the base, and more deeply cut, the notches pointed.
FLORES flavi, spicati, pedunculis brevissimis insidentes.	FLOWERS yellow, growing in a spike, and sitting on very short foot-stalks.
CALYX: PERIANTHIUM monophyllum, subrotundum, inflatum, compressum, quadridentatum, dentibus æqualibus, pallide virens, venosum, persistens. *fig.* 1.	CALYX: a PERIANTHIUM of one leaf, roundish, inflated, flattened, having four equal teeth, of a pale green colour, and permanent, *fig.* 1.
COROLLA monopetala, ringens. *Tubus* subcylindraceus, longitudine calycis; *labium superius* galeatum, compressum, emarginatum, margine anteriori utroque violaceo; *labium inferius* trifidum, laciniis lateralibus planis, regulo, intermedia majori, marginibus involutis. *fig.* 2.	COROLLA monopetalous, ringent. *Tube* somewhat cylindrical, the length of the calyx; the upper lip helmet-shaped, flattened, with a notch on the end, front edge bluish on each side, the lower lip trifid, the lateral segments flat and rolled in, the middle one largest, the edges rolled inward, *fig.* 2.
STAMINA: FILAMENTA quatuor, longitudine labii superioris, sub quo recondita, quorum duo breviora, ANTHERÆ incumbentes, basi bifidæ, hirsutæ, *fig.* 3.	STAMINA: four FILAMENTS the length of the upper lip, under which they lie hid, two of which are shorter than the others, ANTHERÆ incumbent, at one end bifid, and hairy, *fig.* 3.
PISTILLUM: GERMEN ovatum, compressum, glabrum, STYLUS filiformis, staminibus longior, STIGMA obtusum, inflexum, *fig.* 4.	PISTILLUM: GERMEN ovate, flattened, smooth, STYLE filiform, longer than the stamina, STIGMA blunt, and bent downwards, *fig.* 4.
PERICARPIUM: CAPSULA orbiculata, mucronata, compressa, bilocularis, bivalvis, *fig.* 5.	SEED-VESSEL: a roundish flat CAPSULE of two cavities and two valves, terminating in a short point, *fig.* 5.
SEMINA plurima, majuscula, compressa, subreniformia, libera, *fig.* 6.	SEEDS several, rather large, flattened, somewhat kidney-shaped and free, *fig.* 6.

The seeds of this plant, when ripe, rattle in the husks, and hence its name. LINNÆUS informs us, that this circumstance guides the Swedish peasant in mowing his grass for hay. In the neighbourhood of London hay-making commences while this plant is in full bloom.

It abounds in most of our pastures, and flowers early in June.

Agriculturally considered, we may rank it with the useless plants.

In the third edition of RAY's Synopsis, DILLENIUS, on the authority of Dr. RICHARDSON, adds another species, which he calls *Pedicularis major angustifolia ramosissima flore minore luteo, labello purpureo*. Found near York, and also in Northumberland. This, however, is considered by succeeding Botanists as a variety only, and is not found with us.

Rhinanthus Crista Galli

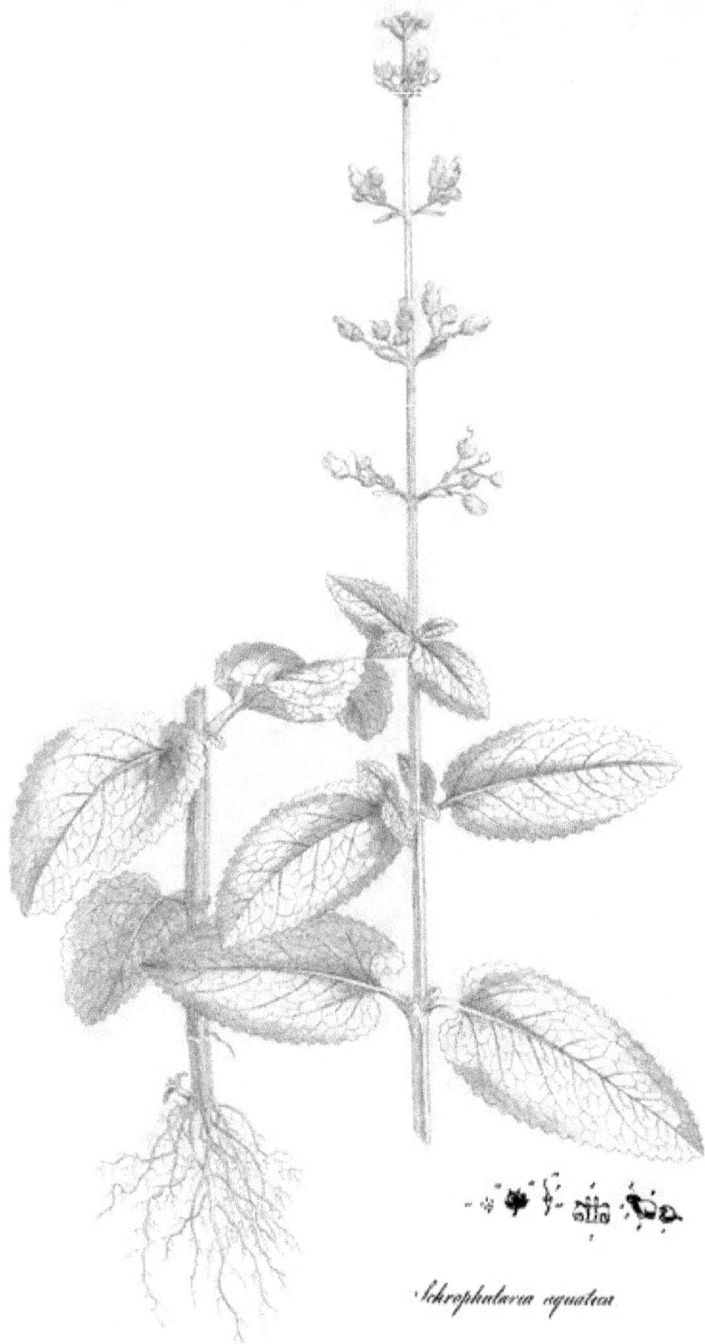

Schrophularia aquatica

Schrophularia aquatica. Water-Figwort, or Water-Betony.

SCHROPHULARIA *Lin. Gen. Pl.* DIDYNAMIA ANGIOSPERMIA.
  Cal. quinquefidus. Cor. subglobosa, resupinata. Caps. biloculacis.
  Raii Syn. Gen. 18. HERBÆ FRUCTU SICCO SINGULARI, FLORE MONOPETALO.
SCHROPHULARIA *aquatica* foliis cordatis obtusis petiolato decurrentibus, caule membranaceis angulato racemis terminalibus, *Lin. Syst. Vegetab.* p. 468. *Sp. Pl.* p. 864.
SCHROPHULARIA caule alato quadrangulo paniculato, foliis ovato lanceolatis. *Hall. Hist.* 326.
SCHROPHULARIA *aquatica. Scopoli Fl. Carn.* n. 776.
SCHROPHULARIA *aquatica major. Bauh. Pin.* 235.
BETONICA *aquatica. Ger. emac.* 715.
BETONICA aquatica major. *Parkinson.* 613. *Raii Syn.* 283. Water-Betony, but more truly Water-Figwort. *Hudson Fl. Angl.* p. 274. *Lightfoot Fl. Scot.* p. 329.

RADIX perennis, crassa, fibris numerosis, majusculis, longis, albis, donata.	ROOT perennial, thick, furnished with numerous, large, long, white fibres.
CAULIS tripedalis, vel orgyalem, erectus, ramosus, lævis, quadrangularis, purpureus, angulis alatis; rami fuffolti, cauli similes.	STALK from three to six feet in height, upright, branched, smooth, four-cornered, purple, the angles winged, branches leafy, like the stalk.
FOLIA petiolata, opposita, distantia, decurrentia, subcoriacea, cordato-oblonga, foliolis appendiculatis, obtusis, venosis, crenata, nuda.	LEAVES standing on foot-stalks, opposite, remote from each other, running in some degree at the base, curved, oblong heart-shaped, having sometimes little appendages, obtuse, veiny, crenated, and smooth.
FLORES paniculato-spicata, terminales.	FLOWERS terminal, growing in a panicle-like spike.
RAMI panicule oppositi, trichotoma, bracteæ ineonoletis suffulti, pedicellis lateralibus, multifloris, inter-terminis, subvifcidis, intermedio solitario.	BRANCHES of the panicle opposite, trichotomous, supported by a pointed bract-leaf, flower-stalks lateral, many-flowered, furnished with floral leaves, somewhat viscid, the middle one solitary.
CALYX: PERIANTHIUM monophyllum, quinquefidum, persistens, laciniis lateralibus, rotundatis, membranä fufcä lacerä marginatis, *fig.* 1.	CALYX: a PERIANTHIUM of one leaf, divided into five segments and permanent, the segments shorter than the corolla, round and edged with a ragged brown membrane. *fig.* 1.
COROLLA monopetala, inæqualis, atro-rubens. Tubus globosus, magnus, inflatus, *fig.* 2. Limbus quinquepartitus, laciniis duabus emissioribus subcrectis, rotundatis, *fig.* 3. cum intercedula squamula laterum parvum mentientes subjecta, *fig.* 4. duabus lateralibus patulis, *fig.* 5. tertia minima tubinvoluta, *fig.* 6.	COROLLA monopetalous, unequal, of a deep red colour. Tube globular, large inflated. *fig.* 2. Limb deeply divided into five segments the two uppermost of which are largest, somewhat upright, and rounded, *fig.* 3. with an intermediate little scale like a small lip placed underneath them, *fig.* 4. the two side ones spreading, *fig.* 5. the third very minute and rolled up, *fig.* 6.
STAMINA: FILAMENTA quatuor, alba, linearia, subvifcida, declinata, longitudine corollæ, quorum duo inferiora. ANTHERÆ didymæ, flavæ, *fig.* 7, 8.	STAMINA: four white, linear, slightly viscid FILAMENTS, inclining downwards, the length of the corolla, two of which are later than the others. ANTHERÆ double and yellow, *fig.* 7, 8.
PISTILLUM: GERMEN subconicum, glandula nectarifera cinctum, *fig.* 9, 10. STYLUS subobtusus, apice subincurvatus, *fig.* 11. STIGMA obtusum, flavum, *fig.* 12.	PISTILLUM: GERMEN subconical, conical, supported by a nectareous gland, *fig.* 9, 10. STYLE tapering, bending downwards a little at the top, *fig.* 11. STIGMA blunt and yellow, *fig.* 12.
PERICARPIUM: CAPSULA subrotunda, acuminata, bilocularis, bivalvis, dissepimento a marginibus valvularum inflexis constructo, apice dehiscens, *fig.* 13.	SEED-VESSEL a roundish pointed CAPSULE, of two cavities and two valves, partition formed by the edges of the valves turning in, opening at top.
SEMINA plurima parva, fusca.	SEEDS numerous, small, and brown.
RECEPTACULUM ovum, subrotundum in utrumque loculamentum se infinuans.	RECEPTACLE single, roundish, insinuating itself into each cavity or cell.

The name of *Water-Betony* (by which this plant is, perhaps, more generally better known than by its other name of *Water-Figwort*) has been assigned it from the great similitude which its leaves bear to those of the *Wood-Betony*; but as it differs from it totally in its fructification, and consequently in its generic character, the latter name is certainly to be preferred.

In its usual state of growth it has little to recommend it as an ornamental plant; but when variegated, few exceed it in beauty. In that state it is not uncommon in the nurseries about London.

It grows naturally by the sides of rivers, ponds, and wet ditches; and flowers from *June* to *September*.

Medicinally the leaves of this species are recommended for the same purposes of those of the *Scrophularia nodosa*, to which they have by some been preferred: in taste and smell they are similar, but weaker. Mr. MARCHANT reports, in the Memoires of the French Academy, that this plant is the same with the *Ipeca* of the Brasilians, celebrated as a specific corrective of the ill flavour of Sena. On his authority the Edinburgh College, in their common infusion of that drug, directed two-thirds its weight of the Water-figwort leaves to be joined : but as they have now disused this ingredient, we may presume that it was not found to be of much use. *Lewis's Mat. Med. Ed. Edin.* p. 598.

The disagreeable smell which attends this plant when bruised makes it rejected by cattle in general; nevertheless, both its leaves and flowers are much relished by different kinds of insects. The *Teuthredo Scrophulariæ Lin.* feeds on its foliage, both in its caterpillar and perfect state. The beautiful caterpillar of the *Phalæna Verbasci* feeds on this plant as well as on the Mulleins. Both bees and wasps collect great quantities of honey from its flowers, and as these continue to be produced for a great length of time, it is one of those plants which perhaps may be made to grow near bee-hives with advantage.

Thlaspi campestre.

THLASPI CAMPESTRE. MITHRIDATE MUSTARD.

THLASPI *Lin. Gen. Pl.* TETRADYNAMIA SILICULOSA.

    Silicula emarginata, obcordata, polysperma: valvulis navicularibus, marginato-carinatis.

    Raii Syn. Gen. 21. HERBÆ TETRAPETALÆ SILIQUOSÆ ET SILICULOSÆ.

THLASPI *campestre* siliculis subrotundis, foliis sagittatis dentatis, incanis. *Lin. Sp. Pl.* p. 902. *Syst. Vegetab.* p. 491. *Fl. Suec. n.* 575.

NASTURTIUM foliis imis petiolatis ovatis, caulinis sagittatis dentatis. *Haller. Hist.* n. 509.

THLASPI campestre. *Scopoli Flor. Carn.* n. 827.

THLASPI arvense, Vaccariæ folio majus. *Bauh. Pin.* 106.

THLASPI mithridaticum sive vulgatissimum Vaccariæ folio. *Parkins.* p. 835.

THLASPI vulgatius. *J. Bauh. II.* p. 921.

THLASPI vulgatissimum. *Ger. em.* p. 262. *Raii Syn.* 305. Mithridate Mustard, Bastard Cresses. *Hudson Fl. Angl.* p. 281. *Lightfoot Fl. Scot.* p. 341.

RADIX annua, simplex, fibrosa.

CAULIS pedalis ad sesquipedalem, erectus, teres, subangulatus, villosus, superne tantum ramosus.

FOLIA radicalia longe petiolata, oblongo-ovata, obtusa, siepius foliaceo, interdum vero subpilosulus, cito marcescentia, caulina sagittata, sparsa, conferta, suberecta, villosa, dentata, amplexicaulia.

FLORES minimi, albi.

RACEMI longi, erecti.

PEDUNCULI teretes, villosi, patentes, siliculis paulo longiores.

CALYX: PERIANTHIUM tetraphyllum, foliolis ovatis, obtusis, concavis, ad latitem flexibilis, marginibus et apicibus albidis, alterutro paulo breviorum et angustiorum, *fig.* 1.

COROLLA: PETALA quatuor, alba, calyce paulo longiora, limbo subrotundo, ungue gracili, *fig.* 2.

STAMINA: FILAMENTA sex, quorum duo paulo breviora. ANTHERÆ flavæ, *fig.* 3.

PISTILLUM: GERMEN ovale, compressum, emarginatum. STYLUS brevissimus. STIGMA capitatum, *fig.* 4.

PERICARPIUM: SILICULA ovata, obtusa, emarginata, disperma, inferne gibba, superne concava, seminibus proeminentibus, *fig.* 5. 6.

ROOT annual, simple, and fibrous.

STALK a foot or a foot and a half high, upright, round, very slightly angular, villous, branched at top only.

LEAVES next the root standing on long foot-stalks, of an oblong ovate shape, for the most part nearly entire, but sometimes pinnatifid at the base, soon decaying, those of the stalk arrow-shaped, placed irregularly, numerous, nearly upright, villous, toothed, and embracing the stalk.

FLOWERS very small and white.

RACEMI long and upright.

FLOWER-STALKS round, villous and spreading, a little longer than the seed-pods.

CALYX: a PERIANTHIUM of four leaves, the leaflets ovate, obtuse, hollow, slightly hairy when magnified, the edges and tips whitish, the alternate ones shorter and narrower than the others, *fig.* 1.

COROLLA: composed of four white PETALS, a little longer than the calyx, the limb roundish, and claw very slender, *fig.* 2.

STAMINA: six FILAMENTS, of which two are shorter than the rest, *fig.* 3.

PISTILLUM: GERMEN oval, flat, emarginate. STYLE very short. STIGMA forming a little head, *fig.* 4.

SEED-VESSEL: an ovate POD, obtuse, emarginate, containing two seeds, underneath globose, above concave, the seeds proeminating, *fig.* 5. 6.

The *Thlaspi arvense siliquis latis* of C. Bauhine, and the present species, are the two whose seeds have been selected from this numerous genus for medicinal use. These appear to have been used indiscriminately; and sometimes the seeds of the common Cress (*Lepidium sativum*) have been substituted for both. Their virtues appear to be nearly similar: RUPPIUS prefers those of the *arvense*, as being the most active; they certainly have much more of the acrimonious taste than those of the *campestre*.

In the present practice they are rarely made use of any otherwise than as ingredients in the Venice Treacle and Mithridate, though some recommend them in different disorders, preferably to the common Mustard, with which they agree nearly in their pharmaceutic properties. *Lewis, Mat. Med.* p. 647.

The present species is not an annual inhabitant of corn-fields; nevertheless it is rather a scarce plant with us. We have noticed it in the gravelly plenty about Church-Wood, near Edgeston. Dr. Goodenough informs me, it is not uncommon in Gunnersbury Lane, near Ealing.

It flowers in June, and ripens its seeds in July and August.

SINAPIS ALBA. WHITE MUSTARD.

SINAPIS *Lin. Gen. Pl.* TETRADYNAMIA SILIQUOSA.

Cal. pateus. *Cor.* ungues recti. *Glandula* inter stamina breviora et pistillum, interque longiora et calycem.

Raii Syn. Gen. 21. HERBÆ TETRAPETALÆ SILIQUOSÆ ET SILICULOSÆ.

SINAPIS alba, filiquis hispidis: rostro obliquo longissimo ensiformi. *Lin. Syst. Vegetab. p.* 503. *Sp. Pl. p.* 933. *Haller Hist.* 466.

SINAPIS alba. *Scopoli Fl. Carn. n.* 843.

SINAPI agit foliis. *Bauh. Pin.* 99.

SINAPI album filiqua hirfuta, femine albo vel rufo. *J. B. II.* 856.

SINAPI fylvestre minus? *Parkins.* 830. *Raii Syn. p.* 295. White Mustard. *Hudson. Fl. Angl. ed.* 2. *p.* 298. *Lightfoot Fl. Scot. p.* 361.

RADIX annua, simplex, fibrosa, albida.

CAULIS sesquipedalis ad bipedalem, erectus, ramosus, crassiusculus, striatus, tener, fragilis, hirsutus, pilis retrorsum, rigidiusculis, deorsum versis.

FOLIA petiolata, alterna, radicalia et pleræque caulina, pallide virentia, venosa, utrinque hirtiuscula, genuina trium circiter parium, inferioribus minoribus, caulina lobulibus, omnibus varie dentata.

FLORES lutei, terminales.
PEDUNCULI tetragono-striati.
CALYX: PERIANTHIUM tetraphyllum, foliolis patentibus, concavis, deciduis, lævibus, sublinearibus, apice obtufis, *fig.* 1, 2.
COROLLA: PETALA quatuor, subrotunda, plana, patentia, integra, unguibus erectis, linearibus, longitudine vix calycis, *fig.* 3.
STAMINA: FILAMENTA sex, quorum duo breviora, viridefcentia, fubulata. ANTHERÆ luteæ, erectæ, fubfagittatæ, *fig.* 4.

GLANDULÆ ut in plerifque hujus generis, *fig.* 5.
PISTILLUM: GERMEN obovatum, fubangulofum, ad lentem hispidum. STYLUS infidetur, anceps, geminæ dorfo five longior, ftaminibus paulo brevior. STIGMA capitatum, *fig.* 6.

PERICARPIUM: SILIQUA hirfuta, fubarticulata, fubtetrafperma, rostro longissimo ensiformi coronata, *fig.* 7, 8.
SEMINA majuscula, fulca, *fig.* 9.

ROOT annual, fimple, fibrous, and whitish.

STALK a foot and a half to two feet high, upright, branched, fomewhat clumfy, finely grooved, tender, brittle, and hirfute, the hairs numerous, ftiffifh, and turned downward.

LEAVES ftanding on foot-ftalks, alternate, thofe next the root and moft of thofe on the ftalk pinnated, of a pale green colour, veiny, flightly hirfute on both fides, compofed of three or four pair of pinnæ, the lowermoft of which are very fmall, the terminal one often three-lobed, and all of them varioufly indented.

FLOWERS yellow, and terminal.
FLOWER-STALKS having four grooves or corners.
CALYX: a PERIANTHIUM of four leaves, which are fpreading, concave, deciduous, fmooth, fomewhat linear, and blunt at top, *fig.* 1, 2.
COROLLA: four roundifh PETALS, flat, fpreading, entire, above upright, linear, fcarcely the length of the calyx, *fig.* 3.
STAMINA: fix FILAMENTS, two of which are fhorter than the reft, of a greenifh colour, and tapering. ANTHERÆ yellow, upright, fomewhat arrow-fhaped, *fig.* 4.

GLANDS as in moft of this genus, *fig.* 5.
PISTILLUM: GERMEN inverfely ovate, flightly angular, hifpid when magnified. STYLE tapering, two-edged, almoft twice the length of the germen, and a little fhorter than the ftamina. STIGMA forming a little head, *fig.* 6.

SEED-VESSEL: a hairy Pod, fomewhat jointed, containing about four feeds, terminated by a very long fword-fhaped beak, *fig.* 7, 8.
SEEDS rather large and brown, *fig.* 9.

In the corn-fields in Buckinghamfhire, efpecially about High Wycomb, the *Sinapis alba* is as common, and as troublefome a weed among the corn as the *arvenfis* with us is found more fparingly. It is frequently met with on banks, and among the corn in Barton-fields, and well known to conftitute a part of young fallading.

RAY has been particularly happy in pointing out the ftriking characters of the feveral fpecies of *Sinapis*, which LINNÆUS has adopted. The feed-veffels, either in their form, fize, or manner of growth, will always with certainty diftinguifh them; but as thefe parts may occur when they are not fufficiently advanced to exhibit thofe characters, it is neceffary to call in others to our affiftance: we may then, in addition to LINNÆUS's fpecific characters, obferve, that the *Sinapis alba* is moft obvioufly diftinguifhed from the *nigra* by having its filiks finely grooved, and from the *arvenfis*, and from the *orientalis*, for which it is perhaps much more liable to be miftaken, by having its leaves more divided to fuggeft as our figure expreffes.

It flowers in June, and ripens its feeds in July.

Sinapis alba

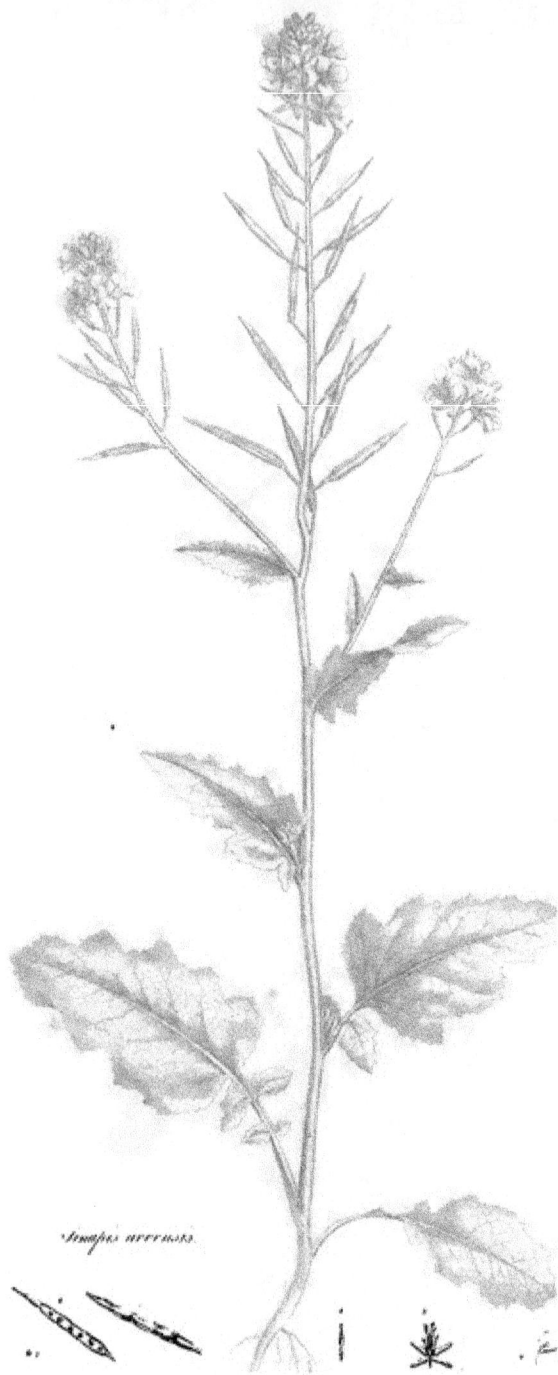

Sinapis arvensis.

SINAPIS ARVENSIS. CHARLOCK.

SINAPIS *Lin. Gen. Pl.* TETRADYNAMIA SILIQUOSA.

> *Cal. patens. Cor. ungues recti. Glandula inter stamina breviora et pistillum, interque longiora et calycem.*

Raii Syst. Gen. 15. HERBÆ TETRAPETALÆ SILIQUOSÆ ET SILICULOSÆ.

SINAPIS *arvensis* siliquis multangulis torulo-turgidis lævibus rostro *acipiti longioribus. Lin. Syst. Vegetal.* p. 503. *Sp. Plant.* p. 933. *Fl. Suec.* 610. *Haller. Hist. n.* 487.

SINAPIS *arvensis. Scopoli Fl. Carn. n.* 841.

RAPISTRUM flore luteo. *Bauh. Pin.* 95.

RAPISTRUM *arvorum. Ger. emac.* 233. *Parkins.* 862. *Raii Syn.* 295. Charlock or Wild Mustard. *Hudson. Fl. Angl.* p. 298. *Lightfoot Fl. Scot.* p. 360.

(Latin)	(English)
RADIX annua, simplex, fibrosa, rigida, albida.	ROOT annual, simple, fibrous, rigid, and whitish.
CAULIS pedalis, sesquipedalis, et ultra, erectus, teres, solidus, striato-sulcatus, hispidus, purpurascens, ramis diffusis.	STALK from one to a foot and a half high, upright, branched, round, solid, striated or grooved, hispid, and purplish, the branches spreading wide.
FOLIA alterna, petiolata, patentia, scabriuscula, venosa, dentato-serrata, ovato-lanceolata, sæpe integra, sæpius vero subsinuata, raro pinnata.	LEAVES alternate, standing on foot-stalks, spreading, roughish, veiny, indented or serrated, ovato-lanceolate, often entire, but most commonly jagged at the base, rarely pinnated.
FLORES lutei, terminales, pedunculati.	FLOWERS of a yellow colour, growing in heads, and standing on flower-stalks.
PEDUNCULI longitudine calycis; hispiduli.	FLOWER-STALKS the length of the calyx, slightly hispid.
CALYX: PERIANTHIUM tetraphyllum, foliolis linearibus, canaliculatis, patentibus, flavis, obtusis, pilosis, *fig.* 1.	CALYX: a PERIANTHIUM of four leaves, the leaves linear, hollowed above, spreading, yellow, blunt and hairy, *fig.* 1.
COROLLA: PETALA quatuor, lutea, obcordata, unguiculata, patentia, unguibus longitudine fere calycis, *fig.* 1.	COROLLA four PETALS of a yellow colour, inversely heart-shaped, spreading, claws almost the length of the calyx, *fig.* 2.
NECTARIA: Glandulæ quatuor saturate virides.	NECTARIES: four Glands of a deep green colour.
STAMINA: FILAMENTA sex, quorum duo breviora, lutea, subulata. ANTHERÆ concolores, incumbentes, primo sagittatæ, apicibus demum revolutis, *fig.* 3.	STAMINA: six FILAMENTS, two of which are shorter than the rest, yellow and tapering. ANTHERÆ of the same colour, incumbent, first arrow-shaped, tips finally rolling back, *fig.* 3.
PISTILLUM: GERMEN cylindraceum, longitudine fere styli, et paulo crassior, nunc læve, nunc hirsutulum. STYLUS longitudine staminum. STIGMA capitatum, bilabiatum, *fig.* 4.	PISTILLUM: GERMEN cylindrical, almost the length of the style, and a little thicker, sometimes smooth, sometimes a little hairy. STYLE the length of the stamina. STIGMA forming a little head, divided into two lips, *fig.* 4.
PERICARPIUM: SILIQUA teres, vix angulosa, patens, lævis aut hirsuta, polysperma, rostro brevi subtetragono terminata, *fig.* 5, 6.	SEED-VESSEL: a round Pod, scarce perceptibly angular, spreading, smooth or hirsute, containing many seeds, terminated by a short somewhat four-cornered beak, *fig.* 5, 6.
SEMINA plurima, minuta, nigricantia.	SEEDS numerous, minute, and blackish.

There are three plants peculiar to corn-fields, which, in various parts of the kingdom, are more or less common, and all of which are apt indiscriminately to be called CHARLOCK: these are the *Sinapis arvensis, Sinapis albe,* and *Raphanus Raphanistrum;* the first and the last of which are by far the most general. The name of *Charlock* ought, however, to be confined to the *Sinapis arvensis,* the most noxious weed of the three, and as such most carefully to be extirpated from among the corn.

The leaves of this plant, on their first appearing above ground, and for some time afterwards, resemble those of the turnip so much, that we have known an intelligent farmer deceived by them, and mistaken in his crop. The whole plant, when young, is often eaten by the labouring part of the community; and, like turnip-tops, is no bad substitute to other culinary plants in times of scarcity.

June is the month in which the Charlock flowers most plentifully; but it may frequently be found in blossom earlier, as well as much later. It is not confined to corn-fields, but is almost equally common among rubbish.

It varies much in height, colour of its stalk, number of its branches, and degree of hairiness. Among corn it grows taller, and is less branched. The Stalk, in some situations, is wholly green; but is more frequently purple at the joints, and very often wholly so. The seed-vessels also vary much in colour and hairiness. We have not observed the flowers subject to any variation of colour.

For the means of distinguishing it from the *Raphanus Raphanistrum,* which at first sight it considerably resembles, vid. *Raphanus Raphanistrum* already figured.

Sisymbrium Irio.

SISYMBRIUM IRIO.   LONDON ROCKET.

SISYMBRIUM *Lin. Gen. Pl.* TETRADYNAMIA SILIQUOSA.

Siliqua dehiscens, valvulis rectiusculis.   Calyx patens.   Corolla patens.

Raii Syn. Gen. 21. HERBÆ TETRAPETALÆ SILIQUOSÆ ET SILICULOSÆ.

SISYMBRIUM *Irio* foliis runcinatis dentatis nudis, caule lævi, siliquis erectis.   *Lin. Syst. Vegetab p.* 499. *Sp. Pl.* 921.   *Fl. Suec. n.* 596.

ERYSIMUM latifolium majus glabrum.   *Bauh. Pin.* 101.

IRIS lævis Apulis erucæ folio.   *Col. Ecphr.* 1. 264.

ERYSIMUM latifolium Neapolitanum.   *Park.* 834.   *Raii Syn. p.* 298.   Smoother broad-leaved Hedge-Mustard.   *Hudson. Fl. Angl. ed.* 2 *p.* 297.   *Jacquin. Fl. Austr. tab.* 322.

Tota planta perpetuo glaberrima est, nec ullum pilum aut villum habet, nec simplex sapor geoilum.

RADIX annua, oblitta, exteriori soliditat crassitie, simplex, quandoque ramosa.

CAULIS pedalis, ad bipedalem, teres, hic illic purpurascens, subdur, fronte inferне, nec flexuosus, sæpius ab ipsa basi ramosus.

FOLIA radicalia, quæ brevi marcescunt, et caulina plæraque, sunt pinnatifida, sinuata, inæqualiter dentata aut serrata, petiolata, patentia, flaccida, lobis ut plurimum acutis, extremo magno et longiore, summis hastata, et quædam integerrima ac simplicia.

CORYMBI in racemos producuntur longissimos, modo recti, modo flaccidos.

FLORES pusilli, flavi.

CALYX patens, flavescens, fig. 1.

PETALA obtusa, et oblonga, ungues habent subcrectos, supra hos patentissima, fig. 2.

STAMINA et STYLUS etiam flavescunt, fig. 3. 4.

SILIQUÆ graciles, subteretes, ad femina torulosæ, et bicusules, brevibus insistunt pedunculis et quaquaversum laxe patent, fig. 5.

SEMINA minuta, pallide flaventa, fig. 6.

The whole plant is always perfectly smooth, without any hair or down, having the biting taste of mustard.

ROOT annual, oblong, the thickness of a goose-quill, simple, branched.

STALK from one to two feet high, round, here and there purplish, strong, below rigid, not striated or grooved, often branched quite from the bottom.

LEAVES near the root, which soon wither, and most of those on the stalk are pinnatifid, sinuated, unequally toothed or serrated, standing on foot-stalks, spreading and flaccid, the lobes for the most part pointed, the end one larger and longer, the uppermost leaves hastate, some of them entire and simple.

CORYMBI lengthened out into long racemi, sometimes strait, sometimes flaccid.

FLOWERS small and yellow.

CALYX spreading and yellowish, fig. 1.

PETALS obtuse and oblong, having claws nearly upright, above which they spread widely, fig. 2.

STAMINA and the STYLE are also of a yellowish colour, fig. 3. 4.

PODS slender, nearly round, about two inches long, standing on short foot-stalks, and spreading loosely every way, leads protuberant, fig. 5.

SEEDS minute, of a pale yellow colour, fig. 6.

The *Sisymbrium Irio*, though a scarce plant in most parts of Great Britain, is frequent enough in the neighbourhood of London: we find it on dry banks, especially such as are made of rubbish, walls, and among rubbish in uncultivated places. Its short time of flowering is from July to September. Like many other annuals it is inconstant as to its particular place of growth. In favourable soils and situations it is capable of multiplying itself exceedingly from the great number of seed vessels which it produces. The seeds are very small, and produce only a little through the sides of the fruit-vessel give them the appearance of being pointed rocks a character, which when present will readily distinguish this plant. Mr. Ray observed it at Faulkbourne in Essex, and on the walls of Berwick on the Tweed. That great naturalist remarks, that after the fire of London in the years 1707, 1708, it came up abundantly among the rubbish in the ruins. Morison, who lived at that period, was particularly struck with its singular in appearance, and in his *Preludia Botanica* has a long dialogue on this very subject; in which, whatever laurels he may get as a Botanist, few will think him entitled to any as a Philosopher.

As this book, containing this curious dialogue, is in few hands, we flatter ourselves a copy of it will not be unacceptable to many of our readers.

"*Botan.* Secundo die Septembris, anno Domini, 1666, incepit incendium illud luctuosum et ad solduum, aut quatriduum durasse. Nec opes humanae (divinitus eventu, quum non est malum in civitate, quod non Scrit Dominus) extingui poterat: nisi diabus appetito veutorum cuente (ut ita loquar) negrabat: per triduum aut quatriduum illud. Pub obtomestive spatium, per radere documenturi jugerum, solis aequatuorum, endii perambulandi versus excombitem vetus cœnos. Ante illud tempos; Collegium Gresthamianum dictum tenebant, in vestigio, aedificioret et tectorum, nobis tanta fere objecti copia, Erylimi illius, quod uni levis Apulus alter Fabia Columnis dicitur: Et cœlum marcitem, mentibus doubus per hoc; adeo cunde pullulavit, ut falce quavi Trificum, aut secale demeti potuerit.   *Sec.* Quid inde expeciat, unde provocelle tantam copiam istius Irionis? putas tu; an à fronte seu fatione? *Botan.* Quid quæso, te movet ad tam proportionatam quæstionem, cum nobis in ea cives ambo, Divi Pauli, et illos putabis in medituilio celeberrici Emperei Londini, à mille aut uitren centenis annis. Fuere comburchtae tetlas conservata?   *Sec.* Ergo tanta copia illим torculis, latebat in cellis et cavernis fondis, et folii plurже cæpolis, profluvente.   *Botan.* Unum hoc addam: ego non fuen Plautus, ut ex aliorum reluisioe monstra imponerem; nec Matthaolus ut applongarem ea quae numquam extitisse: sed ut vis appetitos

vesbat

“urbis nec Calumniatoris: meum tibi dicam animum. *Soc.* Dicas quæso? *Botan.* Nullum est semen plantæ, quod producit (confirmatum quam diligentissime) post decennium, perraro post quinquennium: multò minus post centenos aliquot, et mille annos. *Soc.* Ergò aliquis semina istius plantæ, per rudera sparsit. *Botan.* Non credo imò, certò scio tantam istius Irionis, seminis copiam non fuisse in tota insula Britannica, imo nec in Gallia: dubito an in Germania et Italia ipsa: (cujus Neapolis est regnum, ubi frequenter crescebat tempore Fab. Columnæ,) unquam floruit tanta istius plantæ copia, ergò etiamsi seminatores fuissent (ex tua opinione, post hæc tibi à me audita) non poterat tanta copia istius individualis speciei, seminis; à tot Regnis suppeditari. *Soc.* De hoc non multùm tecum dubito: sed quid concludis, sit rationi consentaneum. Unde provenit tanta copia istius Irionis, forte sponte. Sub idem tempus, ibidem vidisti et observasti multas alias plantas, pappescentes, imo gramineas aliasque diversarum classium. *Botan.* Vidi et attentè observavi. *Soc.* Undè hæ aliæ venere? *Botan.* A semine volatili pappescenti quod potest (ut supra clarè satis docui) ad multa Milliaria, vento transferri, et in altum attolli et ubicunque ceciderit, germinat et fructificat. *Soc.* De pappescentibus non dubito quod dicis, insuper Gramina, densè satis proveniunt: in qualibet terra si negligatur; quare non potest tuum Erysimum, seu Irio lævis Apulus alter in ruderibus Londinensibus, sponte etiam provenire. *Botan.* Non est par ratio inter Gramina et Erysimum hoc; Quia Graminum semina spargantur passim; est omnium vegetantium plantarum, in omnibus regionibus, frequentissima et facillime sese propagat. *Soc.* Est planta tamen prolifica, ex supra dictis à te: ergò à semine, multiplicatur. *Botan.* Hoc ego semper credidi, et in hunc horam credo. Unum a te solicitari velim, putasne hanc plantam, Irionem lævem Apulum Col. a quovis hortulano, aut incola hujus civitatis satam, in ruderibus fuisse. *Soc.* Neminem hujus insulæ primò tam curiosum, secundò nec tantæ ejus plantæ seminis, copia instructum fuisse, pro certo ratum et statutum habeo. Quia tam stolidus est nullus furiosus homo, si semina ad manum haberet (quod impossibile supra demonstratum est) ruderibus ducentorum jugerum terræ, solo æquatorum, commisisset. Ergò cum nec à satione, nec à semine, ad aliquot centenos annos in ruderibus latente, produci poterat; hujus plantæ tanta copia. Unde concludere vis, tantam ipsius multitudinem provenisse. *Botan.* Certè et supra dixi ex sale partim volatili, partim fixo, salpetro, sulphure, et ex terra sive calcea aut ruderosa et aqua, mixtaque materia, quocunque modo appelles, per me non stabit. Nescio quid mihi perfectum habere debeo, adhuc. Probabile certè est, hanc plantam tam copiosè provenisse sponte; ut supra dictum fuit. Sed hæc opinio apperit januam ad philosophorum contemplationes, qui indifferenter, credunt cujuslibet generis plantas, arbores, frutices, suffruticesve, ex terra tanquam matrice, sponte sine semine provenire. Sed hæc opinio (ut mihi videtur) repugnat sacræ scripturæ, et rationi. Hæc per dialogum inter nos dicta, impræsentiarum, sat esse puto. Quod reliquum de hac materia; Sociis virtuosis, Parisiensibus, et Londinensibus, viris nobilissimis, clarissimis et doctissimis (ex quorum numero tu etiam ipse es) discutiendum relinquo. Vale, mi doctissime vir.”

SISYMBRIUM *Lin. Gen. Pl.* TETRADYNAMIA SILIQUOSA.

Siliqua dehiscens, valvulis rectiusculis. *Cal.* patens. *Corolla* patens.

Raii Syn. Gen. HERBÆ TETRAPETALÆ SILIQUOSÆ ET SILICULOSÆ.

SISYMBRIUM *terrestre* radice annua, foliis pinnatifidis dentato-serratis, siliquis fœcundis.

RADIX annua, fibrosa, albida.

ROOT annual, fibrous and whitish.

CAULIS pedalis, sesquipedalis, et ultra, plerumque erectus, ramosus, sulcatus, lævis, viridis, seu purpurascens.

STALK a foot, a foot and a half, or more, in height, generally upright, branched, grooved, smooth, of a green or purplish colour.

FOLIA omnia pinnatifida, Erysimi officinalis quodammodo similia, lævia, pinnis ternis, quatuor, sive sex paribus, cum impari, omnibus inæqualiter dentato-serratis, extima præsertim in inferioribus foliis rotundiora; cætera semiamplexicaulia.

LEAVES, all of them pinnatifid, somewhat like those of Hedge-mustard, smooth, the pinnæ consist of three, four, or six pair, with an odd one, all of them unequally indented, the outermost especially in the bottom leaves roundish, those of the stalk partly amplexicaule.

FLORES minimi, lutei, semper fœcundi.

FLOWERS very small, yellow, and always producing seed.

CALYX: PERIANTHIUM tetraphyllum, foliolis ovatis, obtusis, concavis, subereetis, flavescentibus. *fig. 1. magn.*

CALYX: a PERIANTHIUM of four leaves, which are ovate, obtuse, hollow, nearly upright, and yellowish. *fig. 1. magn.*

COROLLA: PETALA quatuor, lutea, sæpius emarginata, vix longitudine calycis. *fig. 2.*

COROLLA: four PETALS, of a yellow colour, generally nicked at the end, scarcely the length of the calyx. *fig. 2.*

STAMINA: FILAMENTA sex, subæqualia, longitudine pistilli, flavescentia. ANTHERÆ luteæ, incumbentes. *fig. 3.*

STAMINA: six FILAMENTS, nearly equal, the length of the pistillum, of a yellowish colour. ANTHERS yellow and incumbent. *fig. 3.*

PISTILLUM: GERMEN oblongum. STYLUS brevissimus. STIGMA capitatum, villosum. *fig. 4.*

PISTILLUM: GERMEN oblong. STYLE very short. STIGMA forming a little head and villous. *fig. 4.*

PERICARPIUM: SILIQUA teres, longitudine pedunculi, sursum subarcuata, seminibus plurimis haud æqualiter protuberantibus turgida. *fig. 5, 6.*

SEED-VESSEL a round Pod, the length of the flower-stalk, somewhat curved upward, turgid with numerous seeds which protuberate unequally. *fig. 5, 6.*

SEMINA minima, fusca, *fig. 7.*

SEEDS very small and brown. *fig. 7.*

We have taken the name of *terrestre*, which LINNÆUS applies to the third variety of his *Sisymbrium amphibium*, not so much from the certainty of its being the plant he intends, as from the propriety of its application to this species, it being generally found in dryer situations than the true amphibium.

Repeated observation and culture have thoroughly satisfied us that the present plant is a species perfectly distinct from the amphibium; and we ground our authority for considering it as such on the following circumstances.

1st, It is an annual, whereas the amphibium is not only a perennial, but has a creeping root.

2dly, It is a much smaller plant than the amphibium, seldom requiring half its height.

3dly, It is seldom or never found in the water, unless accidentally overflown.

4thly, Its foliage is very different, the radical leaves much resembling those of the Erysimum officinale.

And, lastly, its seed vessels are always turgid, and full of seeds, while those of the amphibium are usually abortive.

As we can find no satisfactory account of this plant either in RAY, HUDSON, LINNÆUS, HALLER, or the numerous authors we have consulted, we have omitted all synonyms, and contented ourselves with giving it a new specific character, chiefly intended to contrast it with the amphibium.

In the course of our botanical researches we have had frequent occasion to remark, that our most common plants are the least known; we seek with avidity such as are rare and with difficulty acquired, and neglect those that we daily tread under foot. The present plant affords an instance of this inattention, as it is a very common one in the environs of London, and found in the same situations as the Rumex maritimus, on the edges of wet ditches, and on ground apt to be occasionally overflown. We have observed it in Tothill-Fields, on the edge of a ditch by the road-side leading from the Magdalen Hospital to Lambeth Marsh, and in our garden it comes up spontaneously as a common weed.

When this plant grows by itself, in a situation tolerably dry, it grows quite erect, and quickly produces a considerable quantity of seeds. Should it happen to be overflown, which is frequently the case, it is then more procumbent, and will sometimes take root at the joints, in which state it appears to be the Sisymbrium palustre repens *portae forte* of VAILLANT, at least it accords in part.

This species of Sisymbrium flowers in June, July, August, and September.

It has a similar taste to most of the plants of the cress kind, but is not very pungent.

Sisymbrium terrestre.

ERYSIMUM OFFICINALE. HEDGE MUSTARD.

ERYSIMUM *Lin. Gen. Pl.* TETRADYNAMIA SILIQUOSA.

Siliqua columnaris, exacte tetraëdra, *Cal.* clausus.

Raii Syst. Gen. 21. HERBÆ TETRAPETALÆ SILIQUOSÆ ET SILICULOSÆ.

ERYSIMUM *officinale* siliquis spicæ adpressis. *Lin. Syst. Vegetab.* p. 499. *Sp. Pl.* p. 922. *Fl. Suec. n.* 598.

ERYSIMUM foliis pinnatis, pinnis rectangulis, acutis, extrema triangulari maxima, siliquis adpressis. *Haller. Hist.* 878.

SISYMBRIUM *officinale. Scopoli Fl. Carn.* n. 824.

ERYSIMUM vulgare. *Bauh. Pin.* 100.

ERYSIMUM Dioscoridis Lobelio. *Ger. em.* 254.

ERYSIMUM vulgare. *Parkins.* 833.

ERUCA hirsuta siliqua caule appressa Erysimum dicta. *Raii Syn.* 298. Common Hedge-mustard. *Hudson. Fl. Angl. ed.* 2. p. 286. *Lightfoot Fl. Scot.* p. 354.

RADIX annuus, descendens, flexuosa, fibrillosa.	ROOT annual, descending, crooked, and fibrous.
CAULIS pedalis ad bipedalem, erectus, teres, striatus, pubescens, scaber, ramosus, sæpius purpurascens.	STALK from one to two feet high, upright, round, finely grooved, beset with numerous short rough hairs, branched, and for the most part purplish.
FOLIA alterna, petiolata, utrinque parciter pubescentia, subtus scabra, præcipue in costis et nervis, pinnatinda, laciniis oppositis, oblongis, serratodentatis, terminali majore, cum laciniis proximis confluente.	LEAVES alternate, standing on foot-stalks, slightly downy on each side, particularly on the mid-rib and nerves, pinnatifid, the segments opposite, oblong, serrated or toothed, the end one largest, and connected with the next to it.
RACEMI florum terminales, foloroturidi; fructuum filiformes, elongati, nudi, pubescentes.	RACEMI of the flowers terminal, roundish; of the fruit filiform, lengthened out, naked, and downy.
CALYX: PERIANTHIUM tetraphyllum, pallidum, foliolis lineari ovalibus, obtusiusculis, concavis, pubescentibus, fig. 2.	CALYX: a PERIANTHIUM of four leaves, of a pale colour, linear-oval, bluntish, concave, and downy, fig. 1.
COROLLA cruciformis, tetrapetala, foedide luteoferous, petalis cuneiformibus, obtusis, venulosis, unguiculatis, calyce longioribus, fig. 4.	COROLLA cross-shaped, composed of four petals, of a dull yellow colour, wedge-shaped, obtuse, veiny, clawed, longer than the calyx, fig. 4.
STAMINA: FILAMENTA sex, subulata, pallida, corolla paulo brevioria; quorum duo adhuc breviora. ANTHERÆ cordatæ, acutæ, subsecuneæ, fig. 2.	STAMINA: six FILAMENTA, tapering, of a pale colour, a little shorter than the corolla; two of which are shorter than the rest. ANTHERÆ heart-shaped, pointed, bent somewhat upward, fig. 2.
NECTARIA: Glandulæ duæ utrinque ad stamina breviora.	NECTARIES: two Glands one on each side, placed at the base of the shorter stamina.
PISTILLUM: GERMEN cylindricum, striatum. STYLUS brevis, pubescens. STIGMA orbiculatum, planiusculum, emarginatum, altitudine fere staminum, fig. 3.	PISTILLUM: GERMEN cylindrical, striated. STYLUS short, downy. STIGMATA round, flattish, emarginate, almost the height of the stamina, fig. 2.
SILIQUÆ cylindricæ, striatæ, virides aut purpureæ, pubescentes, cauli adpressæ, fig. 5, 6.	PODS cylindrical, finely grooved, green or purple, downy and pressed to the stalk, fig. 5, 6.
SEMINA lutesido laetescentia, utrinque oblique truncata, fig. 7.	SEEDS of a dingy yellow colour, obliquely truncated at each end, fig. 7.

The *Erysimum officinale* affords a remarkable instance of that diversity of appearance which the same plant may assume at different periods of its growth. View it just as it comes into blossom, and afterwards, when its flowering branches shoot out horizontally to a great length, and you will scarcely believe that it is one and the same plant.

It grows very commonly on dry banks, under walls, pales, and in waste places; and flowers from June to September.

The leaves of Hedge Mustard are said to be attenuant, expectorant, and diuretic, and stand particularly recommended against chronical coughs and hoarseness, whether temporal or occasioned by immoderate exertion of the voice. Lobel greatly commends for this purpose a compound syrup, which, as Geoffroy observes, is not superior to a simple mixture of the expressed juice of the herb with honey; and indeed it is not very clear, whether the virtue of the honey is much improved by the Erysimum.

The herb has no smell; and its taste, at least when moderately dried, is little other than herbaceous, with somewhat of a slight saline impregnation.

The seeds of Erysimum are considerably pungent, and appear to be nearly of the same quality with those of mustard, but weaker. Their acrimony, like that of mustard-seed, is extracted totally by water, and partially by rectified spirit, and strongly impregnates water in distillation. *Allen's Ed. of Lewis's Mat. Med.* p. 272.

Erysimum officinale.

Lathyrus Aphaca

LATHYRUS APHACA. YELLOW VETCHLING.

LATHYRUS *Lin. Gen. Pl.* DIADELPHIA DECANDRIA.

Stylus planus, supra villofus, superne latior. Cal. laciniæ superiores 2 breviores.

Raii Syn. Gen. 23. HERBÆ FLORE PAPILIONACEO SEU LEGUMINOSÆ.

LATHYRUS *Aphaca pedunculis unifloris, cirrhis aphyllis, ftipulis fagittato-cordatis. Lin. Syft. Vegetab. p. 662. Sp. Pl. 1029.*

LATHYRUS aphyllos ftipulis fagittatis latiffimis. *Haller hift. n. 442.*

LATHYRUS *Aphaca. Scopoli Fl. Carn. n. 887.*

VICIA lutea foliis convolvuli minoris. *Bauh. Pin. 345.*

APHACA *Parkinf. 1067. Ger. emac. 1250. Raii Syn. ed. 3. p. 320. Hudfon Fl. Angl. ed. 2. p. 315.*

RADIX annua, fibrofa.

CAULIS pedalis, fefquipedalis, et ultra, debilis, ope cirrhorum fcandens, tetragonus, lævis.

FOLIA nulla.

STIPULÆ binæ, magnæ, fagittato-cordatæ, obtufæ, utrinque prope bafin denticulo notatæ, glaucæ, fubtus nervofæ.

CIRRHUS fimplex, patens.

FLORES lutei, parvi, folitarii, pedunculati, axillares.

PEDUNCULI foliis longiores, tetragoni, uniflori, bracteâ minimâ prope florem inftructi.

CALYX: PERIANTHIUM monophyllum, quinque partitum, laciniis lanceolatis, fubæqualibus, nervofis, longitudine fere corollæ, fig. 1.

COROLLA papilionacea, VEXILLUM luteum, reflexum, intus lineis cæruleis ftriatum, fig. 2. ALÆ luteæ, fubrotundæ, longitudine carinæ, harum duobus inæqualibus, pallidioribus, fig. 3. CARINA pallide fulphurea, poftice clausa, fig. 4.

STAMINA: FILAMENTA decem, fimplex, et novem fidum, affurgentia, albida, ANTHERÆ fubrotundæ, luteæ, fig. 5.

PISTILLUM GERMEN oblongum, compreffum, viride, glabrum; STYLUS furfum erectus, pallidior, fuperne latior, obtufus; STIGMA a medietate ftyli antice villofus, fig. 6.

PERICARPIUM: LEGUMEN unciale, latiufculum, compreffum.

SEMINA feptem octave, fubrotunda, nitida.

ROOT annual, and fibrous.

STALK a foot, a foot and a half or more in height, weak, climbing by means of its tendrils, four-cornered, and fmooth.

LEAVES none.

STIPULÆ growing in pairs, large, betwixt arrow and heart-fhaped, obtufe, on each fide near the bafe furnifhed with a tooth, glaucous, and ribbed on the underfide.

TENDRIL fimple and fpreading.

FLOWERS yellow, fmall, folitary, growing on footftalks from the ale of the leaves.

FLOWER-STALKS longer than the leaves, four-cornered, one-flowered, furnifhed near the flower with a minute bractea or floral leaf.

CALYX: a PERIANTHIUM of one leaf, deeply divided into five fegments, which are lanceolate, nearly equal, ribbed, and almoft the length of the corolla, fig. 1.

COROLLA papilionaceous, STANDARD yellow, reflexed, ftriped on the infide with blue lines, fig. 2. WINGS yellow, nearly round, the length of the keel, claws two, unequal, paler, fig. 3. KEEL of a pale fulphur colour, clofed behind, fig. 4.

STAMINA: ten FILAMENTS, one fingle, nine connected, rifing upwards, whitifh; ANTHERÆ roundifh and yellow, fig. 5.

PISTILLUM GERMEN oblong, flat, green, and fmooth, STYLE rifing upwards, upright, paler, dilated above, obtufe; STIGMA which rifes from the middle of the ftyle villous on its fore part, fig. 6.

SEED-VESSEL a POD about an inch in length, broadifh, and flattened.

SEEDS feven or eight, roundifh, and fhining.

We have here a very unufual phænomenon in the vegetable œconomy, a plant whofe ftipulæ fupply the place of leaves, at leaft when the plant becomes of a certain age; for, by a kind of accidental examination, we lately difcovered that this fpecies of Lathyrus, foon after it comes up from feed, is ufually furnifhed with one or more pair of leaves, fimilar to the other plants of this family, but which, as the plant advances, totally difappear; thefe are reprefented at *fig. 7.*

A fomewhat fimilar appearance we noticed laft fummer at Mr. MALCOLM'S, *Kennington*, in a rare fpecies of *Mimofa*, called *verticillata*, all the leaves of the young plants were pinnated, and all thofe of the old plants whorled.

LINNÆUS, in his *Species Plant.* takes fome notice of the Aphaca's producing leaves; his words are, *Cirrhus interdum aliquis gerit foliola conjugata, 2, lanceolata, reliquis Lathyris fimillima at hoc rariffime.*

According to our obfervation, the leaves grew on footftalks in the ufual way, without any, or a very fhort tendril, and they were obfervable on every feedling; hence we fufpect them to be common to this plant when young; and rare, merely from being overlooked.

This fpecies is an annual which grows fpontaneoufly in our corn fields, but is not common in the neighbourhood of London; we have obferved it moft frequently about Tottenham and Enfield.

It flowers in June and July.

No particular ufes or noxious qualities are afcribed to it.

Spartium
Scoparium

SPARTIUM SCOPARIUM. COMMON BROOM.

SPARTIUM *Lin. Gen. Pl.* Diadelphia Decandria.

> *Stigma* longitudinale, supra villosum. *Filamenta* germini adhærentia. *Cal.* deorsum productus.

Raii Syn. Arbores et Frutices.

SPARTIUM *Scoparium* foliis ternatis solitariisque ramis inermibus angulatis. *Lin. Syst. Vegetab.* p. 644. *Sp. Pl.* p. 996. *Fl. Suec.* n. 635.

SPARTIUM foliis inferioribus ternatis hirsutis, superioribus simplicibus. *Haller hist.* n. 354.

GENISTA angulosa et scoparia. *Bauh. pin.* 395.

GENISTA cum rapo. *Dodon. Pempt.* p. 761. *Ger. emac.* 1311.

GENISTA vulgaris sive scoparia. *Park. Theat.* p. 228.

GENISTA angulosa trifolia. *J. B. I.* 388. *Raii Syn.* p. 474. Common Broom. *Hudson Fl. Angl. ed. 2.* p. 310. *Lightfoot Fl. Scot.* p. 382.

Frutex tripedalis ad orgyalem et ultra, ramosissimus, ramis erectis, virgatis, viridibus, angulatis, flexilibus, junioribus pubescentibus.

FOLIA sæpius ternata, summis subinde solitariis, foliolis ovatis, **acutis**, pubescentibus, ciliatis, ciliis mollibus inflexis.

PETIOLI pubescentes, complanati.

FLORES lutei, maximi, laxe racemosi.

BRACTEÆ quatuor, obovatæ, inæquales, cruciatæ, obtusæ, ad basin pedunculorum.

PEDUNCULI solitarii, sæpius bini, raro terni, teretes, glabri, stipulis minutis utrasque instructi.

CALYX: Perianthium monophyllum, parvum, bilabiatum, sæpe purpureum, obsolete denticulatum, labiorum apicibus marcidis fuscis, fig. 1.

COROLLA papilionacea, pentapetala, *Vexillum* obcordatum, reflexum, maximum, fig. 2. *Alæ* longitudine carinæ, subovales, breviter petiolatæ, fig. 3. *Carina* amplis et profunda, obtuse rostrata, fig. 4. dipetala, ast in duas partes facile separabilis, margine carinali villis connexo.

STAMINA: Filamenta decem, inferne in unum corpus coalita (hinc decandria non diadelphia) assurgentes, inferioribus longioribus; Antheræ oblongæ, croceæ, fig. 5.

PISTILLUM: Germen oblongum, hirsutum; Stylus subulatus, assurgens, demum spiraliter involutus ad apicem inferne cauſidiſatuſe, Stigma terminale, minimum, capitatum, fig. 6. auct. fig. 7.

PERICARPIUM: Legumen latum, compressum, nigricans, marginibus pilis mollibus ciliatus, fig. 8.

SEMINA plurima ad 20, minuta, subovata, luteſcentia, nitida, fig. 9.

A Shrub from three to six feet high or more, very much branched, the branches upright, twiggy, green, angular, flexible, the young ones downy.

LEAVES most commonly growing by threes, uppermost ones sometimes singly, leaflets ovate, **acute**, downy, edged with soft hairs bending inwards.

LEAF-STALKS downy, flattened.

FLOWERS yellow, very large, growing in loose racemi.

BRACTEÆ four, inversely ovate, unequal, cross-shaped, obtuse at the base of the flower-stalks.

FLOWER-STALKS single, oftener two, rarely three, round, smooth, furnished on each side with a very minute stipula.

CALYX: a Perianthium of one leaf, small, two-lipped, often purple, faintly toothed, extremities of the lips withered and brown, fig. 1.

COROLLA papilionaceous, pentapetalous, *Standard* inversely heart-shaped, reflexed, very large, fig. 2. *Wings* the length of the keel, somewhat oval, on short footstalks, fig. 3. *Keel* large and deep, beak blunt, fig. 4. composed of two petals, or at least easily separated into two parts, the edges being connected together at the keel with soft hairs.

STAMINA: ten Filaments, below united into one body (hence of the class decandria rather than diadelphia) rising upwards, the lowermost ones longest: Antheræ oblong, saffron-coloured, fig. 5.

PISTILLUM: Germen oblong, hairy; Style tapering, rising upward, finally bent spirally, so as to form somewhat more than a circle, near the top hollowed below; Stigma terminal, very small, and forming a little head, fig. 6. magnified, fig. 7.

SEED-VESSEL a broad, flat, blackish Pod, edged with soft hairs, fig. 8.

SEEDS numerous to 20, small, somewhat ovate, dingy yellow, glossy, fig. 9.

The common English Broom is one of the most ornamental shrubs we have, especially that variety of it, in which the calyx is purple, and the blossoms strongly tinged with orange; but even in its common state, such is the profusion of blossoms with which its branches are loaded in the summer, such the charming verdure of its twigs in the winter season, that it may be said to vie with any of the foreign ones, and to be equally deserving a place in all ornamental grounds.

It grows naturally in dry, sandy, barren soils, bears transplanting badly, but is most readily raised from seed.

It is not only in an ornamental point of view, that this plant deserves our notice, it claims our attention also as an useful plant in rural œconomy and medicine.

Though not so commonly used for besoms as the common Heath and Birch, it is preferred for many purposes; in the Northern parts of Great-Britain it is made use of for thatching cottages, corn and hay-ricks, also as a substitute for reeds in making fences or screens; and we have been credibly informed, that in some parts of Scotland, where coals are scarce, whole fields are sown with its seeds to form fuel.

Authors mention the flower-buds, just before they become yellow, as proper for pickling, in the manner of capers * ; the branches, as capable of tanning leather †, and of being manufactured into coarse cloth ‡; the old wood, as furnishing the cabinet-maker with the most beautiful materials for vaneering; and the tender branches, to be frequently mixed with hops for brewing §.

* Dodon, &c. † Haller. ‡ Ibid. § Lightfoot, Fl. Scot.

The

That twigs, when bruised, smell disagreeably; this may, perhaps, be one reason for their being generally rejected by cattle: the plant, however, affords nourishment to a great variety of insects; in particular, to the larvæ of several *Phalænæ* not described by LINNÆUS.

From the roots of this plant springs the Broom Rape, figured in a former number of this work.

" The leaves and stalks of broom have a nauseous bitter taste, which they give out by infusion, both to " water and rectified spirit; and which, on gently inspissating the filtred liquors, remains concentrated in the " extracts: the watery tincture is of a yellowish green or brownish, the spirituous of a dark green colour. " They are accounted laxative, aperient, and diuretic; and in this intention have been often used by the " common people in dropsies and other serous disorders. Dr. MEAD relates a case of an hydropic person, " who, after the paracentesis had been thrice performed, and sundry purgatives and diuretics had been tried " without relief, was perfectly cured, by taking, every morning and evening, half a pint of a decoction of " green broom tops, with a spoonful of whole mustard seed: by this medicine, the thirst was abated, the " belly loosened, and the urinary discharge increased to the quantity of at least five or six pints a day

" Infusions of the ashes of the plant in acidulous wines, have likewise been employed in the same intention, " and often with good success. The virtue of this medicine does not depend, as some have supposed, on " any of the peculiar qualities of the broom remaining in the ashes, but on the alkaline salt and earth, which " are the same in the ashes of broom as in those of other vegetables, combined, wholly or in part, with the " vinous acid. A solution even of the pure earthy part of vegetable ashes, made in vegetable acids, proves " notably purgative and diuretic.

" Of the seeds and flowers, the medicinal qualities are not well known. It is said, that the seeds, in " doses of a dram and a half in substance, and five or six drams in decoction or infusion, prove purgative or " emetic. Some report that the flowers also operate in the same manner; but LOBEL assures us, from his " own observation, that they have been taken in quantity without producing any such effect: and I have " known infusions of the flowery tops drank freely in some asthmatic cases, without any other sensible operation " than a salutary increase of urine and expectoration. The seeds, slightly roasted, are used in some places as " coffee." LEWIS's *Mater. Med.* p. 218.

A variety of this plant, much more hoary than common, is accidentally met with; the most usual time of its flowering with us, is about the latter end of May or beginning of June.

THOMSON, whose observing eye rarely suffered any of the beauties of nature to escape him, has noticed the flowering of this shrub in the following passage, in which he describes the effect which the genial warmth of the season produces on the various animals.

> " While thus the gentle tenants of the shade
> " Indulge their pure loves, the rougher world
> " Of brutes below ruse furious into flame
> " And fierce desire. Thro' all his lusty veins
> " The bull deep-scorch'd, the raging passion feels;
> " Of pasture sick, and negligent of food,
> " Scarce seen, he wades among the yellow broom.

TRIFOLIUM PROCUMBENS. PROCUMBENT TREFOIL.

TRIFOLIUM *Lin. Gen. Pl.* DIADELPHIA DECANDRIA.

Flores subcapitati. *Legumen* vix calyce longius, non dehiscens, deciduum.

Raii Syn. Gen. 24. HERBÆ FLORE PAPILIONACEO SEU LEGUMINOSÆ.

TRIFOLIUM *procumbens* spicis ovalibus imbricatis : vexillis deflexis persistentibus, caulibus procumbentibus. *Linnæi Syst. Veg. p.* 574. *Sp. Pl.* 1088. *Fl. Suec.* n. 673.

TRIFOLIUM spicis serpentibus peuridoris, caulibus erectis. *Haller hist.* 364.

TRIFOLIUM luteum flore lupulino minus. *I. B. II.* 381.

TRIFOLIUM lupulinum alterum minus. *Raii Syn. p.* 330. n. 17. The lesser Hop-Trefoil. *Hudson. Fl. Angl. ed.* 2. *p.* 328. *Lightfoot Flor. Scot. p.* 409.

RADIX annua, fibrosa.	ROOT annual and fibrous.
CAULES plures, spithamæi, pedales et ultra, teretes, durisculi, pilis adpressis pubescentes, præsertim ad extremitates, purpurei, procumbentes, ramosi.	STALKS several, a span, or even a foot or more in length, round, harelish, downy, with hairs pressed close to the stalk, particularly at the extremities, purple, procumbent, and branched.
FOLIA terna, petiolata, remota, inferiora obcordata, superiora obovata, plerumque emarginata, ad apicem argute serrata, plerumque lævia, venis rectis, simplicibus, utrinque impressis.	LEAVES growing three together, remotely, standing on foot-stalks, the lowermost obcordate, the uppermost obovate, for the most part emarginate, towards the top finely serrated, commonly smooth, the veins straight, unbranched, impressed on each side of the leaf.
PETIOLI breves, longitudine stipularum.	LEAF-STALKS short, the length of the stipule.
STIPULÆ binæ, ovatæ, acutæ, quinquenerves, ad margines pilosæ, basi amplexicaules.	STIPULÆ growing in pairs, ovate, pointed, five-ribbed, edged with hairs, and at the base embracing the stalk.
PEDUNCULI unciales circiter, pubescentes.	FLOWER-STALKS about an inch in length and downy.
SPICÆ subrotundæ, multifloræ (raro infra octo, aut ultra viginti) laxius imbricatæ.	SPIKES roundish, many flowered, flowers seldom fewer than eight or more than twenty, loosely imbricated.
FLORES parvi, lutei, pedicellis brevissimis, insidentes.	FLOWERS small and yellow, sitting on very short foot-stalks.
CALYX : PERIANTHIUM quinquedentatum, persistens, subpilosum, dentibus tribus inferioribus longioribus, subulatis, *fig.* 1.	CALYX : a PERIANTHIUM with five teeth, permanent, and somewhat hairy, the three lowermost longer than the rest, and awl-shaped, *fig.* 1.
COROLLA papilionacea, persistens, marcescens, demum rufa, venis saturatioribus striata, *fig.* 2.	COROLLA papilionaceous, permanent, and withering, finally becoming of a reddish brown colour, and striped with veins of a deeper colour, *fig.* 2.
PERICARPIUM : LEGUMEN ovatum, compressum, monospermum, deorsum reflexum, corollâ persistente inclusum, *fig.* 3.	SEED-VESSEL an ovate, flat Pod, turning backward, inclosed in the corolla, which continues, and containing one seed, *fig.* 3.

The *Trifolium procumbens* is often found larger, but more frequently much smaller, than the specimen we have here figured. When it grows luxuriantly it bears a near resemblance to the *agrarium* already published : but in that species the spikes are not only much larger, but also much more closely imbricated, compared with the *procumbens* the *agrarium* may be considered with us at least as a scarce plant; while that is found only in certain spots, the *procumbens* is met with every where, there being scarcely a dry, hilly pasture, or grass plot, on which it may not be found. In its dwarf state it comes very near to the *filiforme* figured in *Ray's Synopsis, tab.* 14. *fig.* 4. indeed it is very difficult to assign their respective folia ; but both Mr. HUDSON and Mr. LIGHTFOOT agree in making the *filiforme* a distinct species ; and the latter assures us, that culture proves them to be specifically different.

All the Trefoils are considered as affording excellent pasturing and fodder for cattle. The present species is, perhaps, not inferior to any of them in these respects ; but the quantity it affords is so trifling, that it can scarcely be thought worth cultivating, especially as it is only an annual.

It flowers during the greatest part of the summer.

HALLER describes it as growing upright, which it never does with us, unless drawn up by surrounding herbage.

Peziza parvonetior.

VICIA CRACCA. TUFTED VETCH.

VICIA *Lin. Gen. Pl.* Diadelphia Decandria.

Stigma future inferiore transverse barbatum.

Raii Syn. Gen. 23. Herbæ flore papilionaceo seu leguminosæ.

VICIA *Cracca* pedunculis multifloris, floribus imbricatis, foliolis lanceolatis pubescentibus, stipulis integris. *Lin. Syst. Vegetab. p.* 553. *Sp. Pl. p.* 1035. *Fl. Suec. n.* 652.

VICIA foliis lanceolatis ferietis, racemis multifloris reflexis, stipulis integerrimis. *Haller. Hist. n.* 424.

VICIA *Cracca. Scopoli Fl. Carn. n.* 899.

VICIA multiflora. *Bauh. Pin.* 345.

VICIA multiflora seu spicata. *Park.* 1072.

CRACCA. *Riv. Tetr.* 49. *Raii Syn. p.* 322. Tufted Vetches. *Hudson. Fl. Angl. p.* 317. *Lightfoot Fl. Scot. p.* 394.

RADIX perennis, repens.
ROOT perennial and creeping.

CAULIS bipedalis, tripedalis et ultra, pro ratione loci, scandens, angulato-fulcatus, pubescens, fragilis, frangendo crepitans, ramosus.
STALK two, three feet or more in height, according to its place of growth, climbing, angular, grooved, downy, brittle, snapping when broken, branched.

STIPULÆ binæ, semifagittatæ, integræ aut denutæ.
STIPULÆ growing in pairs, each resembling half an arrow, entire, or toothed.

FOLIA pinnata, pinnarum 8 seu 12 pariûm, raro ultra, oblongo-lanceolata, mucronata, utrinque sericea pube albida, pinnis oppositis alternatis, cirrho tripartito terminata.
LEAVES pinnated, composed of 8 or 12 pair, seldom more, oblong, lanceolate, terminated by a point, covered on each side with a kind of white silky down, the pinnæ opposite or alternate, terminated by a tripartite cirrhus.

FLORES racemosi.
FLOWERS growing in bunches or racemi.

RACEMI alterni, multiflori, primo suberecti, apice incurvi, postea reflexi, flosculis 10 ad 40, violacei, conferti, breviffime pedicellatis.
RACEMI alternate, many flowered, at first nearly upright, with the tip bent in, afterwards reflexed, flowers from 10 to 40, of a violet colour, crouded together, and standing on very short foot-stalks.

CALYX: Perianthium monophyllum, tubulatum, coloratum, quinquedentatum, dentibus tribus inferioribus longioribus, plolio, modo productionc, duobus superioribus minimis, fig. 2.
CALYX: a Perianthium of one leaf, tubular, coloured, having five teeth, the three lowermost longer than the upper ones, the middle one farthest extended, the two upper ones very minute, fig. 2.

COROLLA: Vexillum emarginatum, reflexum, violaceum, venis faturatioribus obsolete striatum. Alæ connitentes. Carina albida, ad apicem maculâ saturate violaceâ, utrinque notatum, fig. 1.
COROLLA: Standard emarginate, reflexed, of a violet colour, faintly striped with veins of a deeper colour. Wings closing. Keel whitish, marked on each side at the tip with a deeply violet-coloured spot, fig. 1.

STAMINA: Filamenta 10, simplex et novem sidera, alia. Antheræ parvæ, luteæ.
STAMINA: ten Filaments, nine united, one single, white. Antheræ small and yellow.

GERMEN oblongum, compressum, glabrum. Stylus subereflus, undique pilosus. Stigma obtusum, fig. 3.
GERMEN oblong, compressed, smooth. Style nearly upright, hairy all round. Stigma blunt, fig. 3.

PERICARPIUM: Legumen femunciale, pallide fuscum, glabrum, utrinque compressum, fig. 4.
SEED-VESSEL: a Pod about half an inch long, of a pale brown colour, flattened on each side, fig. 4.

SEMINA quatuor vel quinque in singulis leguminibus subrotunda, nigricantia, fig. 5.
SEEDS four or five in each pod, nearly round and blackish, fig. 5.

Linnæus, Haller, and Scopoli, afcribe to this plant *stipulæ integræ*. Indeed the two former found a part of their specific character on this very circumstance; but this character is certainly a very fallacious one, as the plant is frequently found with us having *stipulæ dentatæ*, and such is the specimen we have figured. It has, however, other characters by which it is obviously distinguished. The most striking are drawn from the leaves and flowers: the former are covered with a fine kind of silky down, which gives them a manifest whiteness. This is most apparent in such specimens as grow in dry, exposed situations. The flowers are of a rich deep purple colour, grow in long bunches or racemi, thickly crouded together, and are confpicuous at a distance.

It is a very common plant in the neighbourhood of London, and no where more plentiful than in Battersea Meadows. When it has an opportunity of climbing up a hedge, it will grow to the height of five or six feet; and it is then that its blossoms are displayed to advantage. In the open pastures and fields, it is found much more dwarfish.

It flowers from July to September.

Gentlemen who wish to decorate the hedges of their plantations cannot select a more proper plant, as it is not apt, like the great Bindweed, Travellersjoy, and other strong growing plants, to fuffocate the shrubs which support it.

It is recommended also, by some authors, as affording excellent fodder for cattle.

Vicia Cracca.

Crepis tectorum.

CREPIS *Lin. Gen. Pl.* SYNGENESIA POLYGAMIA ÆQUALIS.

Recept. nudum. *Cal.* calyculatus, squamis deciduis. *Pappus* plumosus, stipitæus.

Raii Syn. Gen. 6. HERBÆ FLORE COMPOSITO, NATURA PLENO LACTESCENTES.

CREPIS *tectorum* foliis lanceolato-runcinatis sessilibus lævibus, interioribus dentatis. *Lin. Syst. Vegetab.* p. 600. *Sp. Pl.* p. 1133. *Fl. Suec. n.* 705.

HEDYPNOIS *tectorum* caule folioso ramoso, foliis runcinatis mollis, radicalibus lanceolatis, caulinis sagittatis acutis sessilibus. *Hudson. Fl. Angl. ed.* 2. p. 341.

CREPIS foliis ad terram pinnatis, superne amplexicaulibus pinnatis habitata. *Haller. Hist. n.* 31.

CREPIS *tectorum*. *Scopoli Fl. Carn. n.* 954.

HIERACIUM luteum glabrum sive minus hirsutum. *I. B. II.* 1024.

CICHOREUM pratense luteum lævius. *Bauh. Pin.* 126. *Park.* 778.

HIERACIUM aphacoides. *Ger. em.* 297.

HIERACIUM foliis et facie chondrillæ. *Parkins.* 794. *Raii Syn.* p. 165. Smooth Succory Hawkweed. *Lightfoot Fl. Scot.* p. 446.

RADIX annua, simplex, potam fibrosa, descendens, lutescens.

ROOT annual, simple, furnished with few fibres, descending, yellowish.

CAULIS pedalis, bipedalis et ultra, erectus, angulatostriatus, nunc glaber, nunc hirsutulus, præsertim inferne, sæpe purpureus, foliosus, ramosus.

STALK from one to two feet high or more, upright, somewhat angular and finely grooved, sometimes perfectly smooth, sometimes a little hairy, especially towards the base, often purple, leafy, and branched.

FOLIA valde variabilia, sæpe tota glabra, alias utrinque hirsutula, radicalia runcinata perfectisis, sed paulo angustiora, nervo medio superne purpureo, caulina amplexicaulia, acuta, varie dentata, ea mea subintegra, linearia, subsagittata, marginibus revolutis.

LEAVES extremely variable, sometimes perfectly smooth, sometimes slightly hirsute on both sides, those near the root very like the leaves of dandelion, but a little narrower, the midrib purple on the upper side, those of the stalk embracing the stalk, pointed, and variously indented, those of the branches nearly entire, linear and somewhat arrow-shaped, the edges rolled back.

FLORES inter minores hujus familiæ, flavi, læve corymbosi.

FLOWERS smaller than most of this family, yellow, and growing loosely in a kind of corymbus.

CALYX communis duplex, exterior brevissimus, patulus, interior subcylindraceus, simplex, sulcatus, squamis erectis, linearibus, coarctantibus, æqualibus, longitudinaliter pilis globuliferis hispidulis, squamæ ad basin quinque aut plures, subulatæ, breves, inæquales, laxæ, pariter hispidulæ.

CALYX common to all the florets double, the exterior one very short and spreading, the interior one somewhat cylindrical, simple, and grooved, the scales upright, linear, contracting, equal, longitudinally beset with stiff hairs, having a little globule at their extremities, the scales at the base are about five or more in number, subulate, short, unequal, loose, and like the others slightly hispid.

COROLLA composita, imbricata. *Corollulæ* hermaphroditæ, plurimæ, æquales, propria monopetala, truncata, quinquedentata, subtus plurumque purpurea, *fig.* 1.

COROLLA compound, and imbricated. *Florets* hermaphrodite, numerous and equal, each single floret monopetalous, truncated, having five teeth, and for the most part purple beneath, *fig.* 1.

STAMINA: FILAMENTA quinque, capillaria, brevissima. ANTHERÆ cylindraceæ, tubulosæ, *fig.* 2.

STAMINA: five, very short, capillary FILAMENTS. ANTHERÆ united into a cylindrical tube, *fig.* 2.

PISTILLUM: GERMEN inferum. STYLUS filiformis, longitudine staminum. STIGMATA duo, reflexa, *fig.* 3.

PISTILLUM: GERMEN somewhat ovate. STYLE filiform, the length of the stamina. STIGMATA two, turned back, *fig.* 3.

SEMINA viginti et ultra in singulo capitulo, fusca, striata; *Pappus* semine longior, sessilis, simplex, *fig.* 4.

SEEDS twenty or more in each head, brown, and finely grooved; *Down* longer than the seed, sessile, and simple, *fig.* 4.

The great variety of appearances to which this plant is subject, in common with many others of the same class, has occasioned no small confusion among botanists, especially the older ones, who have divided it into several species: even modern botanists, and those of the first character, have confessed the difficulty of distinguishing it in its various states. LINNÆUS exclaims, *Nulla planta hac vulgatior, nulla magis fructura et facie varians, nulla magis confusis synonymis.* HALLER writes, *Inseparabilus tenebra synonyma obducunt;* and SCOPOLI says, *Medius Deveret: Crepis* VARIA.

Perhaps nothing short of repeated observation will enable a botanist to distinguish the same plant in its various states, especially such as are subject to such unusual variations; yet there is frequently some character not liable to be altered by difference of soil and situation, which, if pointed out, will be of great service in directing those who may not have plants constantly before them. RAY observes, that the flowers, heads, and seeds of this plant are smaller than those of any other English Hawkweed, the *Hieracia* excepted (he might have added the *Hypochæris glabra*). To the smallness of the flowers, &c. may be joined the structure of the calyx and the stem-clasping leaves; and when it is known to be a plant growing generally in this country on dry banks, in pastures, and on walls, we flatter ourselves there will be little difficulty, with the assistance of our figure, which represents the plant of its medium size, in distinguishing it at all times.

It flowers from June to September.

Mr. HUDSON has thought proper to remove it from the genus *Crepis* of LINNÆUS, with which it must be owned it does not well accord, and make it an *Hedypnois*; yet it does not very well agree with the character he himself has given of that genus; for the pappus can scarcely be said to be subplumosen, unless very highly magnified.

Scorzonera hispanica

LEONTODON HISPIDUM.   ROUGH DANDELION.

LEONTODON *Lin. Gen. Pl.* SYNGENESIA POLYGAMIA ÆQUALIS.

Recept. nudum.   Calyx imbricatus, squamis laxiusculis.   Pappus plumosus.

Raii Syn. Gen. 6. HERBÆ FLORE COMPOSITO, NATURA PLENO LACTESCENTES.

LEONTODON *hispidum* calyce toto erecto, foliis dentatis integerrimis hispidis : setis succeatis.   *Lin. Syst. Vegetab. p.* 596.   *Sp. Pl.* 1124.   *Fl. Suec. n.* 694.

HEDYPNOIS scapo nudo unifloro, foliis lanceolatis dentatis hispidis.   *Hudson Fl. Angl.* 342.

PICRIS caule nudo, unifloro, foliis asperis dentatis.   *Haller. Hist. n.* 25.

LEONTODON *hispidum.*   *Scopoli Fl. Carn. n.* 977.

TARAXACONOIDES perennis et vulgaris.   *Vaill. Act.* 1721. *p.* 232.

HIERACIUM asperum folio magno dentis leonis.   *Bauh. Pin.* 127.

HIERACIUM dentis leonis folio hirsutum.   *Ger. em.* 303.

HIERACIUM asperum foliis et floribus dentis leonis bulbosa.   *Park.* 788.

DENS LEONIS hirsutus λεπτόσαυκος. Hieracium dictus.   *Raii Syn. p.* 171.   Rough Dandelion commonly called Dandelion Hawkweed.   *Lightfoot Fl. Scot. p.* 433.

RADIX perennis, obliqua, e nigro-fusca, plurimis fibris pallidioribus, in terram recte demissis capillata.	ROOT perennial, oblique, of a blackish brown colour, furnished with numerous fibres of a paler colour, running straight into the earth.
SCAPI plerumque plures ex eadem radice, pedalis aut sesquipedalis, erecti, teretes, fistulosi, hirsuti, simplices, nudi, subinde foliolo five pluribus instructi, superne obvie striati et incrassati, ad basin purpurei.	STALKS usually several from the same root, a foot or a foot and a half high, upright, round, hollow, hirsute, simple, naked, now and then furnished with one or more small leaves, above obliquely striated and thickened, purple at the base.
FOLIA radicalia plurima, in pratis suberecta, in apricis supra terram expansa, palmaria seu spithamaea, petiolata, oblonga, sinuato-dentata, obtusiuscula, pallide viridis, hirsuta, pilis ut etiam scapi furcatis.	LEAVES: radical leaves numerous, in meadows nearly upright, in exposed situations expanded on the ground, a hand's breadth or more in length, standing on foot-stalks, oblong, indented and toothed, bluntish, of a pale green colour, hirsute, the hairs as also those of the stalk forked at the extremity.
FLORES majusculi, lutei, ante florescentiam semper nutantes.	FLOWERS largish, yellow, before blowing always drooping.
CALYX sordide virens, squamae laxe imbricatae, inaequales, pilis longis albidis plerumque simplicibus hirsutae.	CALYX of a dingy green colour, scales loosely imbricated, unequal, rough with long whitish hairs, which are for the most part simple.
COROLLA composita, aequalis, flosculis quinquedentati, tubus superne pilosus *fig.* 2.	COROLLA compound, equal, florets furnished with five teeth, the tube hairy on the upper part, *fig.* 2.
SEMINA oblonga, sublinearia, longitudine fere pappi, exteriores paululum incurvati, interiores recti, ad lentum transverse rugosi, *fig.* 3.	SEEDS oblong, nearly linear, almost the length of the pappus, the outer ones bending a little inward, the innermost ones straight, when magnified transversely wrinkled, *fig.* 3.
PAPPUS pilosus, sessilis, *fig.* 4.	DOWN hairy, and sessile, *fig.* 4.
RECEPTACULUM planum, nudum, punctatum.	RECEPTACLE flat, naked and dotted.

Like the other plants of the class *Syngenesia*, the *Leontodon hispidus* is subject to vary considerably in size and hairiness; but very luckily it has one character which attends it in all its states, and which never fails to distinguish it, *its blossoms droop while in the bud*: striking as this character is, we believe it has escaped the observation of former Botanists, at least it has not been considered as of the first consequence in ascertaining the species. The bluntness of its stalks also contributes to distinguish it from some other plants of the same class, while the hairs on the leaves afford a more minute distinction, being usually bifid, but not always so.

As far as we have had opportunity of observing, it is a very general plant throughout the kingdom, especially where there is chalk or lime-stone. In such sort of pastures it abounds as much as the common Dandelion does in rich cultivated ones, and when in flower, which is usually in July, clothes them in the same golden livery.

As it forms so considerable a part of our pasturage, it is of some consequence that we should know whether Cattle are fond of it, either fresh or made into hay; and we wished to lay before our readers the result of LINNÆUS or his Pupils experiments on this head; but, though a Swedish plant, it unfortunately proved to be one of those with which no experiments were made.

The common Dandelion, according to the Linnæan character, is certainly no *Leontodon*, the pappus being simple, and Scopoli has accordingly made another genus of it, *Hedypnois*.

Mr. HUDSON has united the present plant, the *Leontodon automnale*, two species of *Crepis*, with the *Picris echioides*, under one genus of the same name *Hedypnois*; and HALLER arranges our plant with his *Picris*. Amidst all this confusion we have thought it best in the present instance to follow LINNÆUS, especially as there is nothing in the fructification of our plant which militates against the generic character of his *Leontodon*.

ONOPORDUM ACANTHIUM. COTTON THISTLE.

ONOPORDUM *Lin. Gen. Pl.* SYNGENESIA POLYGAMIA ÆQUALIS.

Recept. favofum. *Cal.* fquamæ mucronatæ.

Raii Syn. Gen. 9. HERBÆ FLORE EX FLOSCULIS FISTULARIBUS COMPOSITO, SIVE CAPITATÆ.

ONOPORDUM Acanthium calycibus fquarrofis: fquamis patentibus, foliis ovato oblongis finuatis. *Lin. Syst. Vegetab. p.* 607. *Sp. Pl. p.* 1158. *Fl. Suec. n.* 721.

ONOPORDUM caule alato, foliis ovatis dentatis, dentibus angulofis aristatis. *Haller hist. n.* 159.

ACANOS *Spina. Scopoli Fl. Carn. n.* 1013.

SPINA alba tomentofa latifolia fylvestris. *Bauh. pin.* 382.

ACANTHIUM album. *Ger. emac.* 1149.

ACANTHIUM vulgare, *Parkinf.* 1149.

CARDUUS tomentofus, Acanthium dictus vulgaris. *Raii Syn.* 196. Common Cotton Thistle. *Hudfon Fl. Angl. ed. 2. p.* 354. *Lightfoot Fl. Scot. p.* 459.

RADIX biennis.	ROOT biennial.
CAULIS tripedalis ad fexpedalem, ad bafin ufque ramofus, fublanuginofus, per totam longitudinem alatus, alis latis, fpinofis, fpinis lutefcentibus, divergentibus.	STALK from three to fix feet high, branched down to the bottom, fomewhat woolly, winged throughout its whole length, wings broad and fpinous, the fpines yellowifh and diverging.
RAMI longi, diffufi.	BRANCHES long, and fpreading.
FOLIA feffilia, ovata, acuta, decurrentia, finuata, dentata, feu angulofa, utrinque lanugine incana, inferiora amplissima, longitudine fefquipedalia, latitudine fere pedalia, margine fpinofa.	LEAVES feffile, ovate, pointed, running down the ftalk, finuated and indented or angular, covered on both fides with a kind of white woolly down, the lowermoft leaves very large, a foot and a half long, and almost a foot in breadth, fpinous on the edge.
FLORES purpurei, erecti, terminales, magnitudine florum Cardui mariani.	FLOWERS terminal, purple, upright, the fize of thofe of the Milk Thiftle.
CALYX: communis fubrotundus, ventricofus, imbricatus, fquamis numerofis, fpinofis, undique prominentibus, fpinis apice luteis, bafi pilis albis inxertextis, fig. 1.	CALYX: common to all the florets, fomewhat round, bellying out, and imbricated, the fcales numerous, fpinous, projecting on every fide, the fpines yellow at the points, and at the bafe interwoven with white hairs, fig. 1.
COROLLA: compofita, tubulofa, uniformis: Corollæ hermaphroditæ, æquales, monopetalæ, infundibuliformes, tubo tenuiffimo, fig. 2. limbo erecto, ventricofo, quinquefido, laciniis æqualibus, linearibus, fig. 3.	COROLLA compound, tubular, uniform, Florets hermaphrodite, equal, monopetalous and funnel-fhaped, tube very flender, fig. 2. limb upright, bellying out, divided into five equal linear fegments, fig. 3.
STAMINA: Filamenta quinque, capillaria, breviffima: Antheræ purpureæ, in cylindrum coeunt, quinquedentatæ, fig. 4.	STAMINA: five capillary, very fhort Filaments; Antheræ purple, forming a cylindrical tube, terminating above in five teeth, fig. 4.
PISTILLUM: Germen ovatum, fig. 6. Stylus filiformis, ftaminibus longior; Stigma bifidum, fig. 5.	PISTILLUM: Germen ovate, fig. 6. Style filiform, longer than the ftamina; Stigma bifid, fig. 5.
PERICARPIUM nullum, Calyx arcte conniven.	SEED-VESSEL none, the Calyx clofing ftrongly together.
SEMINA obovata, fubcompreffa, obfolete angulata, rugofa, nigrefcentia, fig. 7. Pappus feffilis, ad lentem hifpidulus, fig. 8.	SEEDS inverfely ovate, a little flattened, faintly angular, wrinkled, blackifh, fig. 7. Down feffile, flightly hifpid when magnified, fig. 8.
RECEPTACULUM cellulis membranaceis, tetragonis, ceucolatum, favi inftar, fig. 9.	RECEPTACLE reticulated with fquare, membranous cells, like a honeycomb, fig. 9.

When the Cotton-Thiftle grows to its full fize, in a pure air, uncontaminated by London Smoke, the grandeur and fnowy whitenefs of its foliage render it highly confpicuous and ornamental.

With us it grows moft commonly on the funny fide of dry banks, and occafionally among rubbifh, but very feldom in open fields; hence it proves very little injurious to the hufbandman.

It is diftinguifhed from the Carduus tribe, by having a receptacle fomewhat like a honeycomb, *vid. fig.* 9. It differs alfo in another circumftance. When the flowering is over, the innermoft fcales of the calyx clofe ftrongly together, and preferve the feed; in the Thiftles, as foon as the feed is ripe, the firft hot day opens the heads, expands the pappus, and the leaft wind carries away the feed; in the Onopordum they remain fhut up, and ftrongly defended, nor can they commit themfelves to the earth, or be eaten by birds, till long expofure to the weather has decayed the calyx which enclofes them; on this account, they may afford fuftenance to birds later in the year, when fimilar food is not to be obtained.

June and July are the principal months of its flowering.

It is not very fubject to the depredations of infects, and it is defended by its ftrong fpines from the attacks of moft quadrupeds.

Onopordum Acanthium

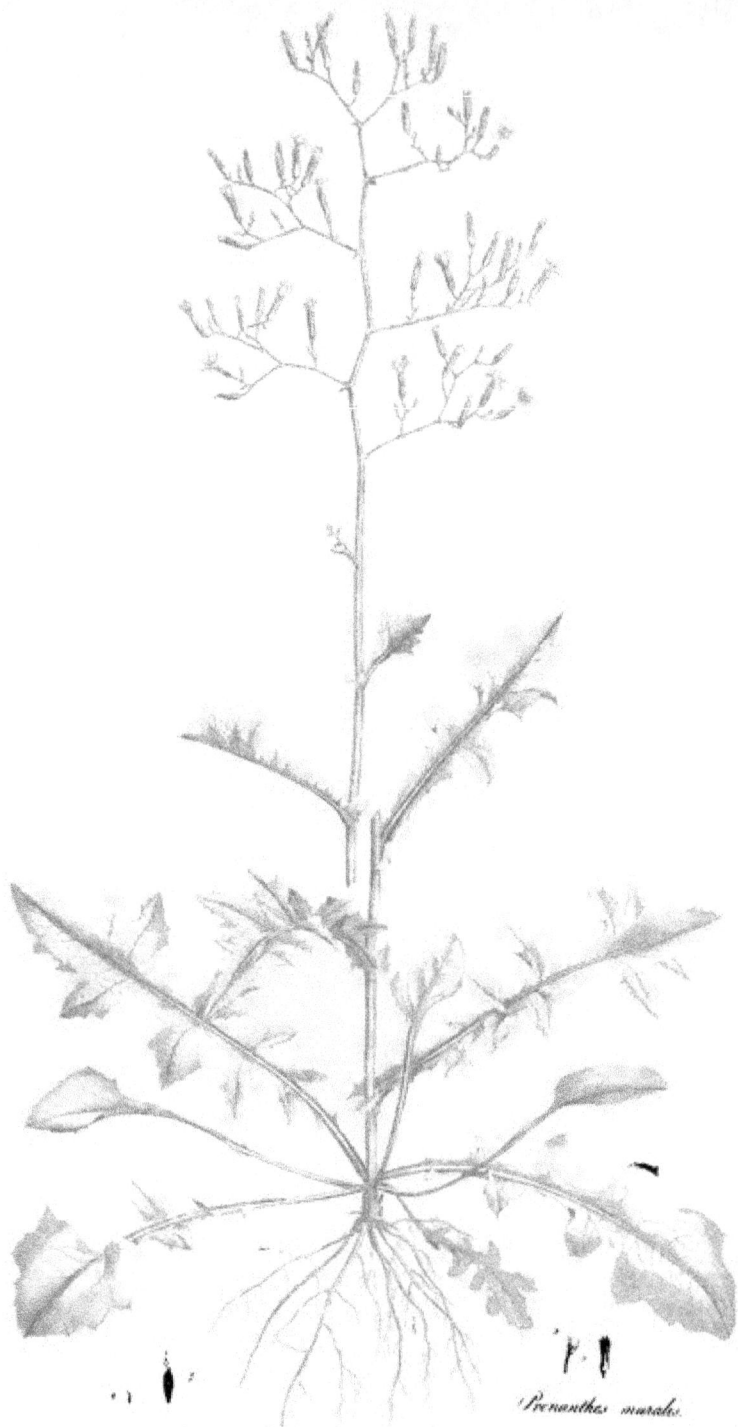

Prenanthes muralis.

PRENANTHES MURALIS. IVY-LEAVED WILD LETTUCE.

PRENANTHES *Linnæi Gen. Pl.* SYNGENESIA POLYGAMIA ÆQUALIS.

Recept. nudum. Calyx calyculatus. Pappus simplex, subsessilis. Flosculi simplici serie.

Rah Syn. Gen. 6. HERBÆ FLORE COMPOSITO, NATURA PLENO LACTESCENTES.

PRENANTHES *muralis* flosculis quinis, foliis runcinatis. *Linn. Syst. Vegetab. p.* 596. *Sp. Pl.* 1121. *Fl. Suec. n.* 692.

PRENANTHES foliis serratis pinnatis, pinna superna triangulari triloboba. *Haller. hist. n.* 18.

PRENANTHES *muralis. Stapnli Fl. Carn. n.* 964.

LACTUCA sylvestris murorum flore luteo. *I. B.* II. 1004.

SONCHUS lævis laciniatus muralis parvis floribus. *Bauhin. Pin.* 124.

SONCHUS lævis muralis. *Ger. emac.* 293.

SONCHUS lævis alter parvis floribus. *Park.* 805. *Raii Syn. p.* 162. Ivy-leaved Sow-thistle, or Wild Lettuce. *Hudson. Fl. Angl. ed.* 2. *p.* 338. *Lightfoot Fl. Scot. p.* 431.

RADIX perennis, ramosa, pallide fusca, lactescens.

CAULIS pedalis ad tripedalem, erectus, simplex, superius subfastuosus, tener, glaucus, purpurascens.

FOLIA radicalia Sonchi oleracei persimilia, inferne purpurea, caulina alterna, amplexicaulia, juncinata.

FLORES parvi, lutei, erecti, paniculati.

PANICULA ampla, nuda, ramosissima, purpurascens.

CALYX communis cylindraceus, glaber, purpurascens, squamis cylindri numerum constellatum, squamis ad basin cylindri triples brevioribus inæqualibus, *fig.* 1.

COROLLA composita, *Corollulæ* hermaphroditæ plerumque quinque, æquales, in orbem simplicem positæ, latiusculæ, nervosæ, quinquedentatæ, *fig.* 2.

STAMINA: FILAMENTA quinque, capillaria, brevissima, flava; ANTHERA æ cylindracea, tubulosæ.

PISTILLUM: GERMEN suboratum; STYLUS filiformis, staminibus longior; STIGMA bifidum, reflexum, *fig.* 3.

SEMEN oblongum, sub acuminatum, nigrum, striatum; PAPPUS breviffimæ petiolatus, simplex, *fig.* 4. basi ausf. *fig.* 5.

ROOT perennial, branched, of a pale brown colour, and milky.

STALK from one to three feet high, upright, simple, leafy, somewhat crooked towards the top, round, glaucous, and purplish.

LEAVES near the root very like those of the common Sow-thistle, purple on the under side, those of the stalk alternate, spreading, and embracing it.

FLOWERS small, yellow, upright, growing in a panicle.

PANICLE large, naked, exceedingly branched, and purplish.

CALYX the common Calyx cylindrical, smooth, purplish, the scales of the cylinder as numerous as the florets, with three, very short, unequal small ones at its base, *fig.* 1.

COROLLA compound, Florets hermaphrodite, usually five in number, equal, forming a single circle, broadish, ribbed, terminated by five teeth, *fig.* 2.

STAMINA: five capillary FILAMENTS, very short and yellow; ANTHERA forming a hollow cylinder.

PISTILLUM: GERMEN suboratum; STYLUS filiform, longer than the Stamina; STIGMA bifid and reflexed, *fig.* 3.

SEED oblong, pointed at the base and striped; Down standing on a very short foot-stalk, simple, *fig.* 4. magnified, *fig.* 5.

Some of the old Botanists considered this plant as a *Lactuca*; others as a *Sonchus*. It approaches nearest to the former, both in its fructification and habit, but yet the foliage is very like that of the *Sonchus oleraceus*. Linnæus, from the paucity of its florets, makes a distinct genus of it, though number seems scarcely sufficient to constitute a generic character. The paucity of florets (there being seldom more than five) at once distinguishes it however from all its kindred; but at the same time we have known it not a little to puzzle students beginning to learn the classes, and who had studied them from such *flowers*, as Dandelion.

It is not a very common plant with us, but is met with occasionally on walls, in woods, and other shady places. We observed plenty of it this year on the outside of the pales which terminate the Terrace in the Spaniard, Hampstead Heath, on the declivity towards Lord Mansfield's little wood.

It flowers from July to September.

SONCHUS PALUSTRIS. MARSH OR TREE SOW-THISTLE.

SONCHUS *Lin. Gen. Pl.* Syngenesia Polygamia æqualis.

Recept. nudum. Calyx imbricatus, ventricosus. *Pappus pilosus.*

Raii Sp. Gen. 27. Herba flore composito, natura [...] Leucocentra.

SONCHUS *palustris* pedunculis calycibusque hispidis subumbellatis, foliis runcinatis basi aristatis. *Lin. Syst. Vegetab.* p. 594. basi *sagittatis. Sp. Pl.* p. 1116.

SONCHUS *asper arborescens. Bauhin. Pin.* p. 124. od. 2.

HIERACIUM *arborescens palustre. Ejusd.* ed. 1.

SONCHUS *tricubitalis*, folio cuspidato. *Morr. Pin.*

SONCHUS *arborescens* alter. *Ger. Em.* p. 294.

SONCHUS lævis altissimus vel Sonchus lævior auctorum 5. altissimus. *Cluf. Hist.* CXLVII.

SONCHUS *arborescens. Parkins.* p. 808. *Raii Syn.* p. 163. The greatest Marsh Tree Sow-thistle: *Hudson. Fl. Anglic.* p. 337.

RADIX perennis, plurimis fibris majusculis capillata, minime vero repens sicut in arvensi.	ROOT perennial, furnished with numerous large fibres, but not creeping, as in the corn Sow-thistle.
CAULIS: ex eodem radice, exsurgunt caules plures, erecti, argylæ, et ultra, crassitie pollicis, angulati, læves, purpurascentes, fistulosi, lacteicentes, folioti, apice ramosi.	STALK: from the same root arise several stalks, upright, six feet or more high, the thickness of one's thumb, angular, smooth, purplish, hollow, milky, and branched at top.
FOLIA caulina sparsa, inferiora basi sagittata, runcinata, laciniis dissimus, vel tribus utrinque inæqualibus, acuminatis, terminali longissimo, supremis integris, cuspidatis, basi sessilia, omnibus minutim denticulatis.	LEAVES of the stalk placed without any regular order, the lower ones arrow shaped at the base, and runcinate, with two or three unequal pointed segments on each side, the terminal one very long, the upper leaves entire, sword-shaped, seated at the base, all of them very finely toothed.
FLORES subumbellati, lutei, floribus arvensis duplo minores.	FLOWERS of a yellow colour, about half the size of those of the corn Sow-thistle, forming a large kind of umbel.
PEDUNCULI hispidi seu potius viscidi caput omnes pili globulo terminantes.	FLOWER-STALKS hispid or rather viscid, as each hair is terminated by a globule.
CALYX constructus primo cylindraceus, apice truncatus, viscidus, persistens florescentia ventricoso-conicus, squamis plurimis, linearibus, inæqualibus.	CALYX: the common calyx at first cylindrical, truncated at top, and viscid, the flowering being over, bellying out at bottom and conical, the scales numerous, linear and unequal.
COROLLA compofita, lacteacens, uniformis. Corollulæ hermaphroditæ, numerosæ, æquales. Tubi longitudine limbi, albus, pilosus. Limbus lineatis, apice quinquedentatus fig. 1, 2.	COROLLA compound, imbricated and uniform. Florets hermaphrodite, numerous, and equal. Tube the length of the limb, white and hairy. Limb lineat, terminated by five teeth, fig. 1, 2.
STAMINA: FILAMENTA quinque, capillata, breviffima. ANTHERÆ flava, in tubum cylindricum coëlita, fig. 3.	STAMINA: five, capillary, very short FILAMENTS. ANTHERÆ yellow, forming a cylindrical tube, fig. 3.
PISTILLUM: GERMEN oblongo-ovatum, album. STYLUS filiformis, longitudine staminum. STIGMATA duo, revoluta, fig. 4, 5.	PISTILLUM: GERMEN oblong-ovate, white. STYLE filiform, the length of the stamens. STIGMATA two, rolled back, fig. 4, 5.
SEMEN pallide fulcum, oblongum, utrinque fulcatum, unde subtetragonum apparet, fig. 6.	SEED pale brown, oblong, with a groove on each side, whence it appears somewhat four cornered, fig. 6.
PAPPUS femine longior, sessilis, simplex.	DOWN longer than the seed, sessile, unbranched.
RECEPTACULUM nudum, punctis prominulis scabrum.	RECEPTACLE naked, rough with small prominent points.

PARKINSON gives a tolerable figure, and a pretty accurate description of this plant; and succeeding Botanists, particularly RAY, have sufficiently ascertained its specific characters: nevertheless HALLER considers it as a variety of the *arvensis*: his words are, " *nec nisi* certis conditionibus *differre videtur.*" Had the Illustrious seen the plant growing, he certainly would not have been thus singular in his opinion.

It agrees with the *arvensis* in having a perennial root, which however does not creep. When placed in a garden, by the side of the *arvensis*, it exceeds it one half: and when planted by the water side, out-tops it by two-thirds. Indeed, in such situations we have seen it ten feet high, and we believe it may justly be considered as the tallest English plant; but though it is so much taller than the *arvensis*, its blossoms are not so large. In its place of growth it differs also from the *arvensis*: while the one is chiefly observed in corn-fields, the other is a constant inhabitant of marshes. There is a difference also in the periods of their flowering, the *palustris* being later by about three weeks; but the base of the leaf in these two plants affords, perhaps, the best character, and of which LINNÆUS, with his usual sharpness, has availed himself.

The *Sonchus palustris* occurs sparingly in the marshes about Blackwall and Poplar, and flowers the latter end of July.

The common Sow-thistle is said to be a favourite food of rabbits; but we believe it has scarcely been suspected, that it might be ranked with our esculent herbs; yet a gentleman, whose delicate state of health has led him to make experiments on such kind of plants, and in whose veracity we place the most implicit confidence, assures us, that he has found the tender shoots and buds of the common Sow-thistle (the smooth sort) boiled in the manner of Spinach, to afford excellent greens, superior to any others which he has tried, not in common use.

Sonchus palustris.

ACHILLEA PTARMICA. SNEEZEWORT.

ACHILLEA *Lin. Gen. Pl.* SYNGENESIA POLYGAMIA SUPERFLUA.

Recept. paleaceum. Pappus nullus. Cal. ovatus, imbricatus. Flosculi radii circiter 4.

Raii Syn. Gen. 8. HERBÆ FLORE COMPOSITO DISCOIDE, SEMINIBUS PAPPO DESTITUTIS corymbosæ DICTÆ.

ACHILLEA Ptarmica foliis lanceolatis acuminatis argute serratis. *Lin. Syst. Vegetab. p.* 647. *Sp. Pl. p.* 1266. *Fl. Suecic.* n. 771.

ACHILLEA foliis linearibus lanceolatis acutissime serratis. *Haller hist.* 117.

DRACUNCULUS serrato folio pratensis. *Bauh. p.* 198.

PTARMICA *Ger. emac.* 605. *Park.* 859. *Raii Syn. p.* 183. Sneezewort, Bastard-Pellitory, Goose-Tongue. *Hudson, Fl. Angl.* 375. *Lightfoot, Fl. Scot. p.* 495.

RADIX perennis, repens, alba, subgeniculata, fibris majusculis et longissimis donata, e geniculis exeuntibus, sapore acri et fervido.

ROOT perennial, creeping, white, somewhat jointed, furnished with large and very long fibres which proceed from the joints, of a hot acrid taste.

CAULIS pedalis ad tripedalem, erectus, plerumque simplex, rigidulus, inferne teres, glaber, superne subangulatus, villosus, paniculatim ramosus.

STALK from one to three feet high, upright, generally simple, somewhat rigid, below round and smooth, above slightly angular, villous, and branching out into a kind of panicle.

FOLIA numerosa, alterna, sessilia, amplexicaulia, linearia, acuta, bi vel tripollicaria, utrinque glabra, inesdiuscula, saturate viridia, margine retrorsum scabra, subcrenata ; cæteris minutim serrulato aculeatis ; subtus trinerviæ, nervis longitudinalibus, quorum intermedius est costa.

LEAVES numerous, alternate, sessile, embracing the stalk, linear, pointed, two or three inches long, smooth on both sides, and somewhat shining, of a deep-green colour, the edge rough, if the finger be drawn along it, from the top to the base, somewhat creased, the another forming a sharp prickly kind of saw, underneath having two longitudinal ribs, beside the middle.

CORYMBUS terminalis, compositus, erectus, villosus, foliosus.

CORYMBUS terminal, compound, upright, villous, and leafy.

BRACTEÆ lineares in pedunculis.

FLORAL-LEAVES linear on the flower-stalks.

CALYX communis hemisphæricus, subtomentosus, imbricatus, squamis ovato-lanceolatis, erectis, subcarinatis, margine rufis, subciliatis.

CALYX common to all the florets, hemispherical, somewhat woolly ; the scales composing it placed one over another, of an oval-pointed shape, upright, somewhat keeled, the margin reddish, and slightly edged with hairs.

COROLLA composita, radiata, flores fœmineis in radio, ligulatis, numero 8-10, laminis ovatis, albis, patentibus, bifidis, apice obtusis, tridentatis, fig. v. tubus marginatus, brevis, longitudine germinis, apice subolata, fig. 1. floras hermaphroditis in disco numerosis, tubo subcylindraceus, marginatus, virescens ; fundo quinquefidus, obtus, tubo breviter, laciniis subrevolutis, fig. 3.

COROLLA compound and radiate, female flowers in the circumference, tubular at bottom and spreading at top, from 8 to 10 in number, the lamina ovate, white, spreading, with two grooves, blunt at top, with three small blunt teeth, fig. 1. the tube two-edged, short, the length of the germen, and reddish at top, fig. 2. hermaphrodite flowers numerous in the centre, the tube nearly cylindrical, two-edged, greenish, the limb white, divided into five segments, shorter than the tube, the segments somewhat rolled back, fig. 3.

STAMINA in hermaphroditis ; FILAMENTA quinque, capillaria ; ANTHERÆ flavæ, in tubum coalitæ, fig. 4.

STAMINA in the hermaphrodite flowers ; FILAMENTS five, very fine ; ANTHERÆ yellow, uniting in a tube, fig. 4.

PISTILLUM in femineis et hermaphroditis : GERMEN compressum, turbinatum ; STYLUS filiformis ; STIGMATA duo, revoluta, apicibus obtusis, fig. 5.

PISTILLUM in the female and hermaphrodite flowers ; GERMEN flattened, broached at top ; STYLE thread-shaped ; STIGMATA two, rolled back, the ends blunt, fig. 5.

SEMINA plurima, nuda, utrinque subalata, nitida, apice truncata.

SEEDS numerous, naked, having a kind of wing on each side, shining, and cut off as it were at top.

RECEPTACULUM paleaceum, squamis membranaceis, lineari-lanceolatis, obtusis, vix longitudine florum.

RECEPTACLE chaffy, the scales membranous, of a shape between linear and lanceolate, blunt, scarcely the length of the flowers.

The dried powder of this plant snuffed up the nostrils provokes sneezing, hence it has acquired its name of *Sneezewort*; chewed in the mouth, like the Pellitory of Spain, it promotes the flow of the saliva, and is deemed serviceable in the cure of the tooth-ach : these appear to be the only medicinal purposes to which it is applied.

In its double state, it has long been an ornament to gardens, and distinguished by the name of *Batchelors Buttons*; having a creeping and very increasing root, it requires more care to destroy than to increase it.

It is a common plant in wet pastures and on heaths, and may be found in plenty by the sides of the ditches in Battersea-Meadows, where it flowers in July and August.

Achillea Ptarmica

ANTHEMIS COTULA. STINKING MAYWEED.

ANTHEMIS *Lin. Gen. Pl.* SYNGENESIA POLYGAMIA SUPERFLUA.

Recept. paleaceum. Pappus nullus. Cal. hemisphæricus, subæqualis. Flosculi radii plures quam 5.

Raii Syn. Gen. 8. HERBÆ FLORE COMPOSITO DISCOIDE SEMINIBUS PAPPO DESTITUTIS CORYMBIFERÆ DICTÆ.

ANTHEMIS *Cotula receptaculis conicis: paleis setaceis, seminibus nudis. Lin. Syst. Vegetab. p. 656. Sp. Pl. p. 1261. Fl. Suec. n. 767.*

CHAMÆMELUM *foliis glabris, duplicato-pinnatis, nervo foliaceo, pinnulis lanceolatis seminibus exasperatis. Haller hift. 103.*

ANTHEMIS *Cotula. Scopoli Fl. Carn. n. 1090.*

CHAMÆMELUM *fœtidum. B. Pin. 135.*

CHAMÆMELUM *fœtidum feu Cotula fœtida J. B. III. 120.*

COTULA *alba Dod. Pempt. 258. Raii Syn. p. 185.* Stinking Mayweed. *Hudson. Fl. Angl. ed. 2. p. 373. Lightfoot Flor. Scot. p. 493.*

Tota planta fœtidiſſima, ſublanuginoſa.	The whole plant extremely fœtid, **and ſlightly woolly.**
RADIX annua, ſimplex, fibroſa.	ROOT annual, ſimple, and fibrous.
CAULIS pedalis ad bipedalem, erectus, ſubangulatus, ſtriatus, pubeſcens, ramoſus, ſæpe uſque ad baſin.	STALK from one to two feet high, upright, ſomewhat angular, finely grooved, downy, branched often almoſt to the bottom.
FOLIA alterna, ſeſſilia, ſublanuginoſa, pinnata, coſtis lineam lata, ſubtus carinata, pleriſque plerumque ramoſis, planis, acutis, ſuperne punctis impreſſis, nudo oculo conſpicuis notata.	LEAVES alternate, ſeſſile, ſlightly woolly, pinnated, the midrib a line broad, keeled underneath, the pinnæ for the moſt part branched, flat, pointed, on the upper ſide marked with impreſſed dots viſible to the naked eye.
PEDUNCULI erecti, ſtriati, nudi, ſuperne ſubincraſſati.	FLOWER STALKS upright, finely grooved, naked, ſomewhat thickened above.
FLORES albi, diſco luteo, minime vireſcente.	FLOWERS white, the centre yellow, without any tendency to green.
CALYX communis, hemiſphæricus, imbricatus, ſquamis pallide virentibus, exterioribus obtuſis, fuſco marginatis, carina ſaturatius virente.	CALYX common to all the florets, hemiſpherical, imbricated, the ſcales of a pale green colour, the outer ones blunt, and edged with brown, the keel more deeply coloured.
FLOSCULI radii tredecim circiter, femellei, ſubovati, lineas duas fere lati, obtuſi, biocavus, tridentati, dentibus obtuſis, fig. 1. pars tubuloſa floſculi ut in Germen, glandulis pellucidis, nudo oculo conſpicuis ornata, fig. 2. Stigma bifidum, laciniis reflexis, ſæpe mancum, fig. 3.	FLOWERS of the radius about thirteen, female, nearly ovate, almoſt two lines broad, obtuſe, two-ribb'd, terminating in three obtuſe teeth, fig. 1. the tubular part of the floret as well as the Germen, ornamented with tranſparent glands, viſible to the naked eye, fig. 2. Stigma bifid, the ſegments reflexed, often imperfect, fig. 3.
FLOSCULI diſci numeroſi, tubuloſi, hermaphroditi, quinquedentati, fig. 4. Stigma bifidum, laciniis revolutis, fig. 6. Germen ut in coſolla ad lentem glanduloſa, fig. 5.	FLOWERS of the diſk numerous, tubular, hermaphrodite, five-tooth'd, fig. 4. Stigma bifid, the ſegments rolled back, fig. 6. Germen as well as the corolla, when magnified, ſtudded with little glands, fig. 5.
SEMEN obtuſe tetragonum, fuſcum, rugoſum, apice planum, punctis in vertice prominulo, excavatis, iiſdem attenuatum, fig. 7. auct.	SEED bluntly four-cornered, brown, wrinkled, flat at top, with a prominent hollow point in the centre, below ſlenderer, fig. 7. magnified.
RECEPTACULUM ſubcylindraceum, ſuperne paleis ſetaceis, rigidis inſtructum, fig. 8.	RECEPTACLE nearly cylindrical, on the upper part furniſhed with rigid, briſtle-ſhaped paleæ or chaff, fig. 8.

The *Anthemis Cotula*, like the *Matricaria Chamomilla*, is very common in corn-fields, where it is well known frequently to bliſter the ſkin of the reapers, or of children who may happen to gather it, which the *Matricaria* never does:—if the plant be examined with a microſcope, it will be found beſprinkled with little glands, in which its acrid matter moſt probably reſides.

Independent of this quality, it abounds to that degree in ſome corn-fields, as greatly to diminiſh the crop. It is fond of a ſoil well manured, and as it is frequently ſuffered to ſeed on dunghills, it by that means often becomes more generally diſſeminated: farmers cannot be too careful in weeding their dunghills; they are not aware of the amazing increaſe from a ſingle plant of the *Anthemis Cotula, Rumex criſpus, Chenopodium album,* or many others equally, if not more, injurious.

We have obſerved the petals to vary much in length and breadth, and Botaniſts have ſometimes found it with double flowers.

It differs greatly in its qualities from the *Anthemis nobilis* and *Matricaria Chamomilla,* but never been much in uſe, nor are its medicinal effects well known. Decoctions of it are ſaid ſometimes to have been employed as a bath or fomentation againſt hyſteric ſuffocation, and hæmorrhoidal pains and ſwellings. Mr. RAY ſays, that a decoction of the herb has by ſome been given internally, with ſucceſs, in ſcrophulous caſes. BROWN LANGRISH gives an account of a decoction of it throwing a perſon afflicted with rheumatiſm into a profuſe ſweat, and curing him. *Lewis's Mat. Med. p. 223. Vid. Matricaria Chamomilla.*

Anthemis Cotula

Chrysanthemum Leucanthemum.

CHRYSANTHEMUM LEUCANTHEMUM. COMMON OX-EYE, or GREATER DAISY.

CHRYSANTHEMUM *Lin. Gen. Pl.* SYNGENESIA POLYGAMIA SUPERFLUA.

Recept. nudum. *Pappus* marginatus. *Cal.* hemisphæricus, imbricatus, squamis marginalibus membranaceis.

Raii Syn. Gen. 8. HERBÆ FLORE COMPOSITO DISCOIDE, SEMINIBUS PAPPO DESTITUTIS, CORYMBIFERÆ DICTÆ.

CHRYSANTHEMUM *Leucanthemum foliis amplexicaulibus oblongis; superne serratis; inferne dentatis. Lin. Syst. Vegetab. ed.* 13 *p.* 772. *Sp. Pl. p.* 1251. *Fl. Suec.* n. 765.

MATRICARIA foliis radicalibus petiolatis, ovatis, crenatis, caulinis amplexicaulibus dentatis. *Haller hist.* 58.

MATRICARIA *Leucanthemum. Scopoli Fl. Carn.* n. 1041.

BELLIS sylvestris caule foliolo major. *Bauh. Pin.* 261.

LEUCANTHEMUM vulgare. *Tourn.* 492.

BELLIS major. *Ger. emac.* 634.

BELLIS major vulgaris five sylvestris. *Parkins.* 528. *Raii Syn. p.* 184. The Greater Daisy, or Ox-Eye. *Lightfoot Fl. Scot. p.* 488. *Hudson. Fl. Angl. ed.* 2. *p.* 371.

RADIX perennis, fusca, subrepens, fibrosa.	ROOT perennial, brown, somewhat creeping, and fibrous.
CAULIS pedalis, sesquipedalis et ultra, erectus, rigidus, angulosus, inferne purpurascens, hirsutus, superne nudus, simplex, subinde ramosus.	STALK a foot or a foot and a half high or more, upright, rigid, angular, below purplish and hairy, above naked, simple, sometimes branched.
FOLIA radicalia a caulinis diversissima, petiolis longis insidentia, obovata, vix pubescentia, inciso-serrata, caulina alterna, sessilia, amplexicaulia, linearia, extrorsum latiora, remote denticulata, denticulis ad basin crebrioribus et longioribus.	LEAVES near the root very different from those of the stalk, standing on long footstalks, obovate, scarcely downy, deeply sawed, those of the stalk alternate, sessile, stem-clasping, linear, outwardly broadest, distantly toothed, teeth at the base more crowded and longest.
FLORES pedunculati, terminales, solitarii, magni, speciosi.	FLOWERS standing on footstalks, terminal, single, large, and showy.
PEDUNCULI striati, subincrassati.	FLOWER-STALKS finely grooved, and somewhat thickened.
CALYX communis hemispherico-planus, arcte imbricatus, squamis exterioribus oblongo-ovatis, obtusiusculis, margine membranaceis, fuscis, interioribus lanceolatis, acutis.	CALYX common to all the florets, like a hemisphere flattened, closely imbricated, exterior scales oblong-ovate, somewhat blunt, the margin membranous and brown, interior scales lanceolate and pointed.
COROLLA composita, radiata; *Discus* luteus, convexus; *Radius* albus patens.	COROLLA compound and radiate; Centre yellow and convex; Circumference white and spreading.
COROLLULÆ *Hermaphroditæ*, tubulosæ, numerosæ, infundibuliformes, quinquefidæ, in disco, *fig.* 1. *Femineæ* 16 circiter, in radio, oblongæ, obtusæ, tridentatæ, *fig.* 5.	FLORETS *Hermaphrodite* tubular, numerous, funnel-shaped, divided into five segments, in the centre, *fig.* 1. Female about 16 in the circumference, oblong, obtuse, three-notch'd, *fig.* 5.
ANTHERÆ fuscæ, in tubum coalitæ, *fig.* 2.	ANTHERÆ yellow, forming a tube, *fig.* 2.
PISTILLUM *Hermaphroditi*: GERMEN oblongum, striatum, angulatum, glabrum, *fig.* 3. STYLUS filiformis, filamentis longior; STIGMATA duo, subrevoluta, superne ad semen canaliculata, apicibus truncatis, crassiusculis, *fig.* 4. *Feminei* GERMEN et STYLUS ut in Hermaphrodito; STIGMA subsimile, laciniis minus revolutis, *fig.* 6.	PISTILLUM of the *Hermaphrodite* flowers: GERMEN oblong, finely grooved, angular, smooth, *fig.* 3. STYLE filiform, longer than the stamina; STIGMATA two, rolled a little back, on the upper part channelled if magnified, the tips truncated and thickish, *fig.* 4. of the *Female* flowers, GERMEN and STYLE as in the Hermaphrodite flowers; STIGMA somewhat similar, but less rolled back, *fig.* 6.
SEMEN oblongum, basi attenuatum, undique profunde sulcatum, ex nigro-purpurascens, *fig.* 7. 8. *fig.* 9. auct.	SEED oblong, slenderer towards the base, deeply grooved all round, and purplish black, *fig.* 7. 8. *fig.* 9. magn.

This species of Chrysanthemum is extremely common in meadows and pastures, sometimes even on walls, and in corn-fields; it is a hardy perennial, increases greatly by seed, and flowers in June and July.

As it is so prevalent in pastures, it is of no small consequence to ascertain how far it is agreeable to cattle, and, on such occasions, the only guide we have at present to consult, are the experiments of LINNÆUS; from these it appears that kine and swine refuse it, but that horses, sheep, and goats feed on it.

The fresh leaves chewed, discover a saccharine, unpleasant, slightly aromatic taste, somewhat like Parsly, but not hot or biting; they have been recommended in disorders of the breast, both asthmatical and phthisical, and as diuretics, but are now seldom called for.

As such a number of beautiful double varieties of the Common Daisy are met with in almost every garden, it has often been matter of wonder to us that we never see this plant in a similar state: I have indeed been very credibly informed, that two double varieties of this plant exist in a garden near Air in Scotland, but never yet saw them.

Matricaria Chamomilla.

MATRICARIA CHAMOMILLA. CORN FEVERFEW, or CAMOMILE.

MATRICARIA *Lin. Gen. Pl.* SYNGENESIA POLYGAMIA SUPERFLUA.

> *Recept.* nudum. *Pappus* nullus. *Cal.* hemiſphæricus, imbricatus: marginalibus ſolidis, acutiuſculis.

Raii Syn. Gen. 8. HERBÆ FLORE COMPOSITO DISCOIDE, SEMINIBUS PAPPO DESTITUTIS, CORYMBIFERÆ DICTÆ.

MATRICARIA *Chamomilla* receptaculis conicis, radiis patentibus, ſquamis calycinis margine æqualibus. *Lin. Syſt. Vegetab.* p. 643. *Sp. Pl.* p. 1256. *Fl. Suec.* n. 764.

MATRICARIA foliis planis capillaribus, duplicato-pinnatis, pinnulis lanceolatis bifidis trifidiſque. *Haller. hiſt.* n. 101.

CHAMÆMELUM vulgare, Leucanthemum Dioſcoridis. *Bauh. pin.* 135.

CHAMÆMELUM *Gerard. emac.* 754.

CHAMÆMELUM vulgare *Parkinſ.* 85. (qui vulgare cum nobili confundit) *Raii Syn.* p. 185. *Hudſon Fl. Angl. ed.* 2. p. 372. *Lightfoot Fl. Scot.* p. 491.

RADIX annua, ſimplex, fibroſa.	ROOT annual, ſimple, and fibrous.
CAULIS pedalis, ad ſeſquipedalem, erectus, ramoſus, ſubanguloſus, ſtriatus, lævis.	STALK a foot, or a foot and a half high, upright, branched, ſomewhat angular, ſtriated, and ſmooth.
FOLIA ſaturate viridia, alterna, ſeſſilia, lævis, pinnata, pinnis lineatibus, inferioribus ſimplicibus, ſuperioribus ramoſis, pinnulis acutis, mucronatis, divaricatis, coſta ſeſquilineam lata, carinata.	LEAVES of a deep green colour, alternate, ſeſſile, ſmooth, pinnated, the pinnæ linear, the lower ones ſimple, the upper ones branched, the pinnules or ſmall pinnæ ſharp and terminating in a ſharp point, divaricating, the midrib half a line broad, and keeled.
PEDUNCULI erecti, ſtriati, nudi, ſupernè ſubincraſſati.	FLOWER STALKS upright, ſtriated, naked, a little thickened above.
FLORES albi, diſco è luteo-vireſcente.	FLOWERS white, the diſk of a yellowiſh-green colour.
CALYX communis hemiſphæricus, ſquamis plurimis, imbricatis, obtuſiuſculis, apice ſubfuſceſcentibus, ſubmembranaceis, longitudine ferè tubi floſculorum femineorum in radio, *fig.* 1.	CALYX common to all the florets, hemiſpherical, ſcales numerous, imbricated, ſomewhat obtuſe, the tips browniſh, and a little membranous, almoſt the length of the tube of the female flowers in the circumference, *fig.* 1.
FLOSCULI radii 13 numero, feminei, oblongi, ſeſquilineam lati, bifulci, tridentati, dentibus obtuſiuſculis, *fig.* 2. STIGMA bifidum, flavum, laciniis reflexis, *fig.* 3.	FLOWERS of the radius about 13 in number, female, oblong, a line and a half broad, two-grooved, three-toothed, teeth bluntiſh, *fig.* 2. STIGMA bifid, yellow, the ſegments turned back, *fig.* 3.
FLOSCULI diſci, numeroſi, tubuloſi, hermaphroditi, quinquedentati, *fig.* 4. STYLUS bifidum, laciniis reflexis, *fig.* 5.	FLOWERS of the diſk, numerous, tubular, hermaphrodite, five-toothed, *fig.* 4. STIGMA bifid, the ſegments turned back, *fig.* 5.
SEMINA numeroſa, minuta, pallidè fuſca, oblonga, ſulcata, *fig.* 6.	SEEDS numerous, minute, of a pale brown colour, oblong and grooved, *fig.* 6.
RECEPTACULUM oblongum nudum.	RECEPTACLE oblong and naked.

The *Matricaria Chamomilla*, *Anthemis Cotula*, and *Chryſanthemum inodorum*, are three very common plants in the neighbourhood of London; as the two firſt are extremely ſimilar in their general appearance, and are often found growing together, we have publiſhed them in the ſame number, that an opportunity might be afforded of comparing and contraſting them.

PARKINSON, deceived by their great ſimilarity, makes only one plant of them; Maynard, ſays he, is ſo like unto Chamomile, that I muſt needs join them together.

The ſtudent who is acquainted with the mode of inveſtigating the generic character of each, will quickly diſtinguiſh the one from the other; on diſſecting the heads, he will find the pointed paleæ which are fixed to the receptacle of the *Anthemis* totally wanting in the *Matricaria*; but this knowledge, though highly neceſſary, is not ſufficient for thoſe who would wiſh to know plants at firſt ſight, which is always deſirable; we ſhall therefore, in addition to the generic character, point out ſeveral others, in which they have appeared to us materially to differ from each other.

Their place of growth affords but little diſtinction, they are both natives of corn-fields, both grow in them in the greateſt abundance, often together, frequently ſeparate, nor is it unuſual to find them on the confines of dunghills, and by road-ſides; they both flower at the ſame time, from May to July and Auguſt, both are annual, and grow nearly to the ſame height, but in the following particulars they differ: the whole plant in the *Matricaria* puts on a deep green colour, and ſomewhat ſhining appearance; the *Anthemis*, on the contrary, aſſumes a much paler hue, and the ſtalk is often covered with a kind of woolly ſubſtance: the leaves in the *Matricaria* are nearly as fine as thoſe of fennel, which they diſtantly reſemble; in the *Anthemis* they are almoſt twice as broad, and the points of them, which in the *Matricaria* are ſimple, in the *Anthemis* are often bifid.

The Petals in both theſe plants begin to hang down in the evening, and continue to do ſo till morning; but thoſe of the *Anthemis* are in general much broader than thoſe of the *Matricaria*, and ſomewhat ſhorter; but, as this particular, both plants are ſubject to great variation; the diſk of the flower in the *Anthemis* is not ſo prominent, but of a lighter yellow than that of the *Matricaria*. Such are the characters which preſent themſelves to the eye of an accurate obſerver, but there is another which will greatly aſſiſt to corroborate, confirm, and render it impoſſible for the plants to be miſtaken, viz. the ſmell; if the heads of the *Matricaria* are bruiſed, they will be found to emit a ſtrong ſmell, ſomewhat reſembling the true Chamomile, but not ſo pleaſant, while the heads of the *Anthemis*, treated in the ſame manner, ſmell inſufferably diſagreeable; another circumſtance may alſo be added, the *Matricaria* is not known to bliſter the ſkin, in which alone it is perhaps left inſuſceptible to the huſbandman than the other; nor is the character which may be drawn from the ſeeds to be deſpiſed, thoſe of the *Anthemis* being broad and truncated at top, wrinkly, and of a deep brown colour when ripe, thoſe of the *Matricaria* much ſmaller, paler, and differing in their ſhape, vid. *fig.* 6.

July 5th, we obſerved ſeveral larvæ feeding on this ſpecies, which produced the Coſſus *teſtiſx*—Cattle in general refuſe the *Matricaria*—In Sweden the flowers are uſed medicinally inſtead of the *Anthemis nobilis*.

Mr. HUDSON, in our opinion, is perfectly juſtified, in making one plant of the *Matricaria Chamomilla* and *inodorum*; Mr. LIGHTFOOT, in his *Flora Scotica*, previouſly ſuggeſted that they were the ſame. We are ſurpriſed that Profeſſor MURRAY ſhould adopt a ſpecies founded on ſuch vague characters as *radiis deflexis* and *radiis patentibus*.

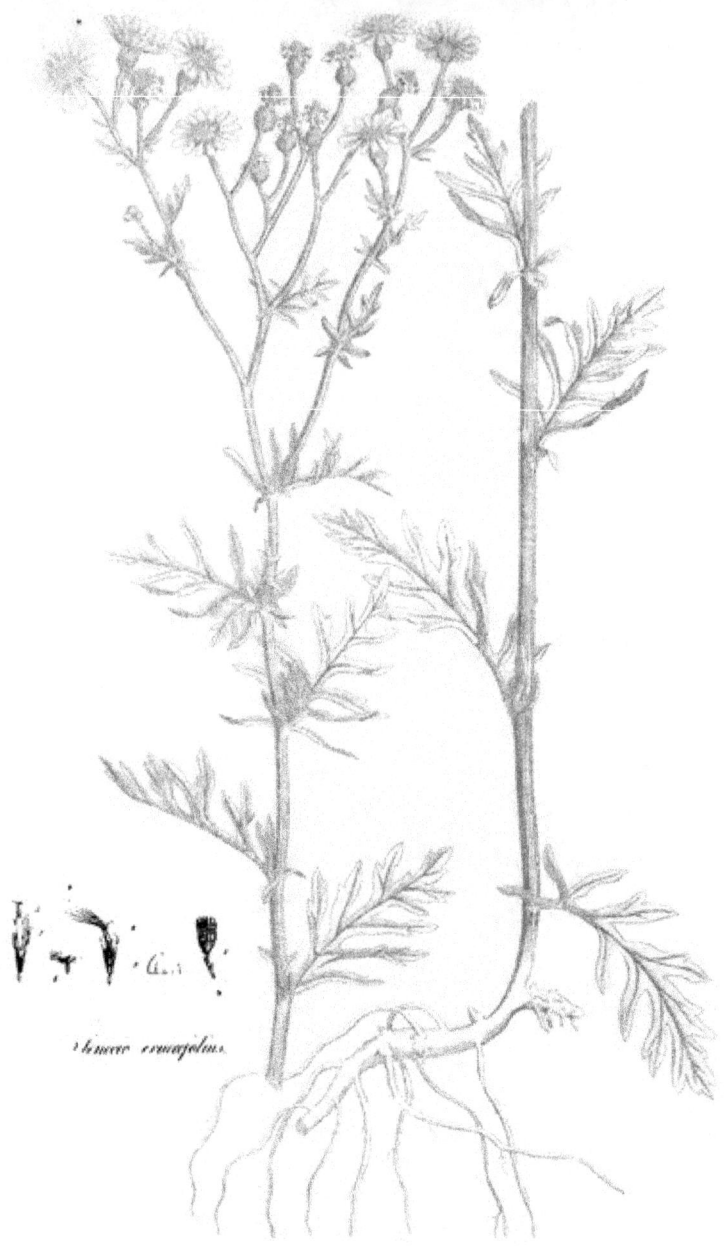

Sisueae eroaefolius.

Senecio erucæfolius. Hoary Ragwort.

SENECIO *Lin. Gen. Pl.* Syngenesia Polygamia Superflua.

> *Recept.* nudum. *Pappus* simplex. *Cal.* cylindricus, calyculatus: squamis apice sphacelatis.

Raii Syn. Gen. 7. Herbæ flore composito, semine papposo non lactescentes flore discoide.

SENECIO *erucæfolius* corollis radiantibus, foliis pinnatifidis dentatis subhirtis, caule erecto. *Lin. Syst. Vegetab.* p. 631. *Sp. Pl.* p. 1218. *H. Suec.* p. 750.

JACOBÆA altissima, foliis erucæ æstatisæve similibus et æmulis. *Rupp. Jen.* 164.

JACOBÆA Senecionis folio incano perennis. *Raii Syn.* p. 177. Hoary perennial Ragwort with Groundsel leaves. *Hudson. Fl. Angl.* p. 366.

RADIX perennis, alba, plures interdum crassitie pennæ anserinæ, a-ociales, aut biunciales, sapore ingrato, in inquentem annum producens.

CAULIS erectus, tripedalis, foliosus, rigidus, subdivisus, purpureus, lævigatus.

FOLIA alterna, semiamplexicaulia, subtus hirsuta, etiam incana, omnia pinnata seu potius pinnatifida, pinnis linearibus, acutis, dentatis.

FLORES lutei, numerosi, corymbosi, magnitudine fere florum Senecionis Jacobææ.

CALYX communis sub-cylindraceus, sulcatus, squamis duodecim, æqualibus, margine membranaceis, apicibus hirsuto-glandulosis, nulla adjeridine floris, squamulis paucis linearibus adpositis ad basin, *fig.* 1.

COROLLA composita, radiata, *Flofculi feminei* in radio tredecim circiter, patentes, oblongi, obfolete tridentati, *fig.* 2. *Hermaphroditi* numerosi in disco, limbo quinquefido, suberecto, *fig.* 3.

STAMINA: Filamenta quinque capillaria. Antheræ in cylindrum coalitæ, *fig.* 5.

SEMEN oblongum, hispidulum, pappo sessili, simplici, subsordidum, *fig.* 6.

ROOT perennial, white, putting forth against the next year several shoots, the thickness of a goose quill, an inch or two inches in length, of a disagreeable taste.

STALK upright, three feet high, leafy, rigid, slightly branched, purple and woolly.

LEAVES alternate, half embracing the stalk, hairy underneath, and sometimes white with down, all of them pinnated, or rather pinnatifid, the pinnæ linear, pointed and jagged.

FLOWERS yellow, numerous, about the size of the flowers of the common Ragwort, growing in a corymbus.

CALYX common to all the florets, somewhat cylindrical, grooved, scales fifteen in number, equal, membranous at the edge, the tips hairy and somewhat glandular, not tinged with a floret, furnished with a few linear scales at the base, which set partial close, *fig.* 1.

COROLLA compound and radiate, *female florets* in the circumference about thirteen in number, spreading, oblong, finely three-toothed, *fig.* 2. *Hermaphrodite flowers* in the centre numerous, the limb divided into five segments and nearly upright, *fig.* 3.

STAMINA, five capillary Filaments. Antheræ united and forming a cylinder, *fig.* 5.

SEED oblong, a little hispid, crowned with sessile, simple down, *fig.* 6.

We have no doubt but the plant here figured is the *Jacobæa Senecionis folio incano perennis* of Ray's *Syn.* ed. 3. p. 177. It certainly has a less jagged, and more groundsel-like leaf, than the common Ragwort. The leaves and stalks are also in general hoary, especially the latter; and so far the description discriminates; but we perceive there both the *apparatus* and *Jacobæa*, with which it has the greatest affinity, so considered as perennial. We believe also, that our plant is the *Jacobæa altissima, foliis Erucæ æstatisæve similibus et æmulis* of Rupp. *Fl. Jen.* Hall. p. 176. And as this latter gives names appears among those which Linnæus enumerates in his *Syst. Veg.* we consider ourselves warranted in adopting the name of *Erucæfolius*. Haller remarks, who can not confidently allow the varieties, that two varieties of species, in the present instance confirms the plant as a variety only of the *Jacobæa*. Prussian Jacquin, in his *Flora Austriaca*, gives a figure and description of a *Senecio*, which he calls *erucæfolius*; but as he adduces no synonyme, and as his figure differs in some respect from our plant, though we rather suspect it to be the same, we dare not consider it as such.

The *Senecio Erucæfolius*, though not so common as the *Jacobæa*, is not unfrequent in the neighbourhood of London in certain situations, particularly in the environs of woods, under hedges, among bushes, &c. and no where more abundant than about the Oak of Honour Wood, near Peckham. The *Jacobæa*, on the contrary, delights to grow in open hilly pastures, church yards, by road sides, &c. where a not of those places either late in their usual period of flowering; the *Erucæfolius* flowering chiefly in August, a month later than the other.

* The hoariness is most observable when the plant is young, or when it grows in a woody and hilly situation, which is its chief abode. When it is found in a moist soil, or cultivated in a garden, it loses this character, in common with many other plants of the same kind.

ORCHIS LATIFOLIA. MARSH ORCHIS.

ORCHIS *Lin. Gen. Pl.* GYNANDRIA DIANDRIA.

Nectarium corniforme pone florem.

Raii Syn. HERBÆ BULBOSIS AFFINES.

ORCHIS *latifolia bulbis subpalmatis rectis, nectarii cornu conico: labio trilobo lateralibus reflexo, bracteis flore longioribus. Lin. Syst. Vegetab. ed.* 14. *p.* 810. *Sp. Pl.* 1334. *Fl. Suec.* n. 801.

ORCHIS *radicibus palmatis, caule fistuloso, bracteis maximis, labello trifido serrato: medio segmento obtuso. Haller. hist.* 1279. *t.* 32.

ORCHIS *latifolia. Scopoli Fl. Carn.* n. 1118.

ORCHIS *palmata pratensis latifolia, longis calcaribus. Bauh. Pin.* 85.

PALMA CHRISTI *mas. Ger. emac.* 220.

ORCHIS *palmata mas f. Palma Christi mas. Park.* 1356.

ORCHIS *palmata non maculata. I. B. II.* 774. *Raii Syn. p.* 380. The Male-Handed Orchis, or Male Satyrion Royal. *Lightfoot Fl. Scot. p.* 516. *Hudson Fl. Angl. ed.* 2. *p.* 385.

RADIX *bulbosa, bulbis palmatis.*

ROOT bulbous, bulbs palmated, or handed.

CAULIS *plerumque pedalis aut sesquipedalis; ad apicem fere foliosus, crassus, fistulosus, superne subangulosus, glaber.*

STALK usually a foot or a foot and a half high, leafy almost to the top, thick, hollow, somewhat angular above, perfectly smooth.

FOLIA *e flavo viridia, suberecta, glabra, nobiscum immaculata, plerisque hujus generis et longiora et latiora.*

LEAVES of a yellowish-green colour, nearly upright, smooth, spotless with us, and both longer and broader than most of this tribe.

FLORES *nobiscum sæpius rosei seu carnei, sæpe purpurei, raro albi, spicati, conferti.*

FLOWERS with us for the most part rose or flesh-coloured, often purple, rarely white, growing in a spike thickly together.

SPICA *subovata, foliosa.*

SPIKE somewhat ovate, and leafy.

BRACTEÆ *magnæ, acuminatæ, coloratæ, fig.* 1.

FLORAL-LEAVES large, long-pointed, and coloured, *fig.* 1.

COROLLA: *petala quinque, duo exteriora ovato-lanceolata, suberecta, parum maculata, fig.* 3. *interiora conniventia, fig.* 4. *Calcar germine brevius, conicum, incurvum, obtusum.*

COROLLA pentapetalous, the two outermost ovato-lanceolate, nearly upright, spotted a little, *fig.* 3. the innermost closing together, *fig.* 4. the Spur shorter than the germen, conical, incurved, and blunt.

NECTARIUM *obsolete trilobum lineolis et punctis saturatioribus pulchre variegatum, lateribus per ætatem reflexis, fig.* 2.

NECTARY faintly three-lob'd, beautifully variegated with small lines and dots of a deeper colour, the sides reflexed with age, *fig.* 2.

STAMINA: FILAMENTA *duo;* ANTHERÆ *subrotundo-clavatæ, e luteo-virescentes, fig.* 5. *auct.*

STAMINA: two FILAMENTS; ANTHERÆ roundish, club-shaped, of a yellowish-green colour, *fig.* 5. magnified.

The *Orchis Latifolia* is particularly distinguished from the others, by growing (with us at least) only in very wet meadows, where *Valeriana dioica, Menyanthes trifoliata,* and *Lychnis Flos Cuculi,* usually abound, and from which circumstance, we have called it *Marsh Orchis,* by its spotless foliage, which is of a yellowish-green colour, and by the uncommon length of the floral leaves, which give the spike a very leafy appearance.

It comes nearest to the *maculata*: HALLER represents the leaves somewhat spotted, and LINNÆUS describes them *parum maculata* ; we do not find them so in the neighbourhood of London ; but probably they may be so in other places: should that be the case, these two plants will approach still nearer to each other.

With us, pink is the most predominant colour of its blossoms, though they are frequently found purple, and sometimes white; even in the same meadow.

We need go no further than Battersea-Meadows to find this plant in tolerable abundance; at a greater distance from town it will be found much more plentifully; it flowers towards the latter end of May.

It is more easily cultivated than many of the same genus, and if planted in a moist border, in a mixture of bog earth and loam, will grow to a much greater size than is represented on the plate.

Orchis latifolia

Sparganium ramosum.

SPARGANIUM RAMOSUM.   GREAT BUR-REED.

SPARGANIUM   *Lin. Gen. Pl.*   MONOECIA TRIANDRIA.

MASC. Amentum subrotundum. Cal. 3-phyllus. Cor. o.

FEM. Amentum subrotundum. Cal. 3-phyllus. Cor. o. Stigma a-fidum. Drupa calucea, 1-sperma.

Raii Syn. GRAMINIFOLIA NON CULMIFERA SINGULARES ET SUI GENERIS.

SPARGANIUM   ramosum foliis basi triangularibus, lateribus concavis, pedunculis ramosis.

SPARGANIUM   erectum foliis erectis triquetris. *Lin. Syst. Vegetab. p. 702. Sp. Pl. p. 1378. Fl. Suec. n. 832.*

SPARGANIUM   caule foliisque erectis. *Haller hist. 1303.*

SPARGANIUM   erectum. *Scopoli Fl. Carn. n. 1146.*

SPARGANIUM   ramosum. *Bauh. Pin. 15. Ger. emac. 45. Parkins. 1205. Raii Syn. 437.* Branched Bur-Reed. *Hudson Fl. Angl. ed. 2. p. 401. Lightfoot Fl. Scot. p. 589.*

RADIX perennis, repens, radiculis fibrillis numerosissimis instructis.	ROOT perennial, and creeping. the small roots furnished with very numerous fibres.
CULMUS bipedalis, tripedalis, et ultra, erectus, teres, glaber, foliosus, folia tribus circiter praeter bracteas.	STALK two, three feet high, or more, upright, round, smooth, leafy. leaves about three in number besides the floral leaves.
FOLIA radicalia erecta, saturate viridia, culmo duplo fere longiora, basi vaginantia, equitantia, paulo supra basin sese ad apicem usque triquetra, latere interiore planiusculo, duobus exterioribus concavis.	LEAVES next the root upright, of a deep green colour, almost twice the length of the stem, sheathy at bottom and riding one on the other, from the base nearly, almost to the top three-cornered, the inner side almost flat, the two outermost hollow.
BRACTEÆ quatuor circiter, foliis caulinis subsimiles, inferioribus longioribus.	FLORAL-LEAVES about four in number, somewhat like the leaves of the stalk, the lowermost longest.
FLORES monoici, in capitula collecti, spicati.	FLOWERS monoecious, formed into little heads, and growing in spikes.
PEDUNCULI axillares, alterni, flexuosi, multiflori, capitulis sessilibus, inferioribus foemineis, duobus aut tribus, superioribus masculis pluribus; pedunculi supremi flores masculos tantum gerunt.	FLOWER-STALKS growing from the bosoms of the leaves, alternate, crooked, supporting many flowers, the little heads sessile, the lowermost ones female, two or three in number, the uppermost ones male, and more numerous; the uppermost flower-stalks bear only male flowers.
CALYX Flor. Masc. Amentum commune, subrotundum, undique densissime imbricatum, constans Perianthiis propriis plerumque triphyllis, basi linearibus, apice ovato-acutis, deciduis, *fig. 1. auct.*	CALYX of the Male Flowers. One common roundish Catkin, closely imbricated on every side, and composed of numerous individual Perianthia, consisting for the most part of three leaves, linear at the base, ovate and pointed at top, and deciduous, *fig. 1. magnified.*
COROLLA nulla.	COROLLA none.
STAMINA: FILAMENTA plerumque tria, capillaria, longitudine calycis; ANTHERÆ oblongæ, flavæ, *fig. 2.*	STAMINA: usually three capillary FILAMENTS, the length of the calyx; ANTHERÆ oblong, yellow, *fig. 2.*
CALYX Flor. Fem. Perianthium ut in masculo, at basi latior, magis concavus, nec deciduus, *fig. 3.*	CALYX of the Female Flowers. A Perianthium as in the males, but broader at the base, more concave, and not deciduous, *fig. 3.*
PISTILLUM: GERMEN oblongo-ovatum, angulatum, definens in STYLUM brevem tubulatum; STIGMA oblongum ad unam latus villosum, *fig. 4.*	PISTILLUM: GERMEN oblong-ovate, angular, terminating in a short tapering STYLE; STIGMA oblong, villous on one side, *fig. 4.*
PERICARPIUM: DRUPA calcarea, turbinata cum acumine, inferne angulata, *fig. 5.*	SEED-VESSEL: a juiceless DRUPE, turban-shaped and pointed; angular below, *fig. 5.*
SEMEN: Nucas dura, ossea, oblongo-ovata, *fig. 6.*	SEEDS: two bony NUTS, of an oblong ovate shape. *fig. 6.*

The *Sparganium ramosum* having a very strong creeping root, is one of those plants which very soon fill up a ditch or piece of water, if suffered to remain unmolested; we have not seen it more plentiful any where than in the Isle of Dogs, the ditches of which are full of it.

We know of no use to which it is applicable.

The stalk is liable to be eaten by some kind of larva whose history we have not yet discovered, the leaves by the larva of a Tenthredo unknown to us, as well as by the larva of the *Phalaena Typhae*—two of which in the r Chrysalis state, we this year, August 24, 1786, found in a web under the leaves of the plant, in a pond near Malden in Essex; and on the leaves of the same plant, at the same time and place, Dr. GOODENOUGH and myself were so fortunate as to find two specimens of that rare insect the *Sphex fusipes. Linnæi.*

The male flowers vary much in the number of their stamina, and both sorts in the number of the leaves of the calyx.

In treating of the *Typha latifolia*, we promised, when we gave a figure of this plant, to inform our readers whether its seeds vegetated: we have since then had an opportunity of observing one of its heads, as it lay on a wet situation, assume a green colour, which, on a careful examination, it was found to owe to the seeds having just begun to vegetate.

SPARGANIUM SIMPLEX. SMALL BUR-REED.

SPARGANIUM *Lin. Gen. Pl.* MONOECIA TRIANDRIA.

  MASC. Amentum fubrotundum. *Cal.* 3-phyllus, *Cor.* o.

  FŒM. Amentum fubrotundum. *Cal.* 3-phyllus. *Cor.* o. *Stigma*
   2-fidum. *Drupa ex fucca,* 1-fperma.

Raii Syn. GRAMINIFOLIÆ NON CULMIFERÆ SINGULARES ET SUI GENERIS.

SPARGANIUM *Simplex* foliis bafi triangulatibus, lateribus planis, pedunculis fimplicibus.

SPARGANIUM fimplex foliis enfiformibus planis, caule fimplici, *Hudfon Fl. Angl. p.* 401.

SPARGANIUM natans foliis decumbentibus planis. *Lin. Syft. Vegetab. p.* 702. *Sp. Pl.* 1378.

SPARGANIUM non ramofum. *Bauh. Pin.* 15.

SPARGANIUM non ramofum. *Parkinf.* 1205. *Raii Syn. p.* 437. n. 2, 3. Bur-reed not branched.

LINNÆUS makes only two fpecies of the genus *Sparganium*, one of which he calls *erectum*, and the other *natans*; the former he defcribes as very common in ditches and fifh-ponds, the latter peculiar to lakes and deep waters.

Older Botanifts defcribe three fpecies, the *ramofum*, the *non ramofum*, and the *minimum*; the *non ramofum* LINNÆUS confiders as a variety of his *erectum*; it is this plant which we here give a figure of, from a thorough conviction of its being a fpecies perfectly diftinct from the common one, whether it differs fpecifically from the *natans* we do not take on us at prefent to determine: Mr. LIGHTFOOT, who has feen the *natans* in many places in Scotland, pronounces it a fpecies; Mr. HUDSON, on the contrary, confiders it as a variety of the prefent plant;—certain it is, foil and fituation will occafion an amazing difference in the appearance of plants; we need only look at the *Polygonum amphibium* to be convinced of this; when it grows on land its leaves are all erect, in the water they float; the leaves of the *Fiftuca fluitans* float in the fpring; as the fummer advances they grow upright; poffibly the depth and confequent coldnefs of the water, with other circumftances, may occafion the prefent plant to affume the floating appearance which authors defcribe;—culture, perhaps, can only decide this matter:—let the experiment turn out as it may, as there are found to be two fpecies with erect leaves, it became neceffary to alter LINNÆUS's names, which Mr. HUDSON having judicioufly done we have adopted them.

We fhall now point out the feveral characters in which the prefent plant has appeared to us to differ from the *ramofum*.

  It differs in its place of growth,
  In its fize,
  In the colour and fhape of its leaves,
  In the branchednefs of its flower-ftalks, and
  In the colour of the male and female flowers.

The common Bur-Reed grows in almoft every ditch in the neighbourhood of London, the fmall one on the contrary is found only in particular fpots, particularly in fuch pools of water as one meets with on heaths, and which are frequently made by the digging of gravel, along with the *Myriophyllum*, the *Alifma Damafonium, Sifon inundatum, Scirpus fluitans,* &c. It particularly abounds on Batterfea Common, juft before you enter Wandfworth on the left-hand fide from London, and flowers during the whole of the fummer.

It is feldom found more than one fourth part fo high as the *Sparganium ramofum*.

The leaves incline much more to a yellow colour, and inftead of being hollow on two fides near the bafe, as thofe of the *ramofum* are, they are flat, fo that a tranfverfe fection forms a triangle with nearly plain fides; we look on this as its beft fpecific character. Such as have opportunities of obferving the *natans*, will do well to obferve whether its leaves are fimilar near the bafe.

Each flower-ftalk fupports only a fingle globule of male or female flowers; the lowermoft which fupport the female flowers vary confiderably in length, being fometimes more than an inch long, and at other times feffile.

The flowers before they blow look yellow, and have none of that blacknefs about them, fo confpicuous in thofe of the *ramofum*: they are alfo larger in proportion.

Sparganium simplex.

The Dogs Mercury was at one period thought to be an innocent plant, its poisonous qualities were discovered by accident: the Annual, or French Mercury has, at present, the reputation of being not only harmless, but to possess medicinal virtues; it is of some consequence then for us rightly to distinguish the two, and in this there is little difficulty. The Dogs Mercury has a strong, creeping, perennial root; this an annual one: the Dogs Mercury flowers only in the Spring; this the whole Summer long: the Dogs Mercury has an unbranched stem; this a stalk branched down to the bottom.

The Annual Mercury has been ranked among the emollient oleraceous herbs; it is said gently to loosen the belly; its principal use has been in glysters.

The whole plant, particularly when in flower, has a strong smell of Elder.

The fine blue colour which the *Dogs Mercury* acquires in drying, has induced several persons to believe, that the plant, if properly treated, might be made, as well as many others, to produce Indigo; this induced Mr. MACINTOSH, an ingenious young gentleman of Glasgow, to make the following chemical analysis of it, with which he was so obliging as to favour me; and though it does not come under the proper plant, we apprehend no apology will be necessary for inserting it here.

" The whole plant, on being put into water, gives out a fine blue colour, which is immediately changed
" into a green by the addition of an alcali; but an acid has not the power of changing its colour into red,
" as it does most blue liquors, it only weakens the blue, and if a large quantity be added, it nearly destroys
" it   The whole plant, on being dried, assumes a blue colour, which it gives out readily to water; but in
" all cafes, if a boiling heat be used, it only acquires a deep dirty green, which changes gradually into a
" brownish red. Upon agitating violently the blue liquor, I always found it was changed into a brown
" colour, the blue being entirely lost, and not to be recovered by any means I could fall upon. There falls
" during this process, a small quantity of precipitate, which is also brown. If the blue liquor be evaporated,
" the whole is likewise changed into the same brownish colour, and a similar precipitate falls, which, on
" being put into water, gives it a dark red colour. Newly-slacked lime put into the blue liquor, first
" changes it into a green, which is very soon after destroyed. I have observed in the beginning of the
" evaporation, a blue secula upon the sides of the vessel, but always before the end of the process, the whole
" was of the brownish colour mentioned above."

MERCURIALIS ANNUA. ANNUAL, or FRENCH MERCURY.

MERCURIALIS *Lin. Gen. Pl.* DIOECIA ENNEANDRIA.

MASC. Cal. 3-partitus. Cor. o. *Stam. 9-12. Antheræ* globosæ didymæ.

FÆM. Cal. 3-partitus. Cor. o. Styli 2. Caps. dicocca, 2-locularis, 1-sperma.

MERCURIALIS annua caule brachiato, foliis glabris, floribus spicatis. *Lin. Syst. Vegetab. p. 746. Spec. Pl. p. 1465.*

MERCURIALIS caule annuo, brachiato, foliis conjugatis, ovato lanceolatis, glabris. *Haller hist. n. 1600.*

MERCURIALIS Cynocrambe *Scopoli Fl. Carn. n. 1226.*

MERCURIALIS testiculata, sive mas Diosc. et Plinii. *Bauhin pin. 121.*

MERCURIALIS spicata, sive foemina, Diosc. et Plinii. *Bauhin pin. 121.*

MERCURIALIS vulgaris mas et femina. *Park. 295.*

MERCURIALIS mas et femina. Ger. emac. 332.

MERCURIALIS annua glabra vulgaris. *Raii Syn. p. 139.* French Mercury, the male and female, *Hudson. Fl. Angl. ed. 2. p. 435.*

RADIX annua, fibrosa, alba.	ROOT annual, fibrous, of a white colour.
CAULIS pedalis ad sesquipedalem, erectus, glaber, ad basin usque ramosus, geniculatus, geniculis incrassatis, subcompressis, utrinque, idque alterne.	STALK a foot or a foot and a half high, upright, smooth, branched quite to the bottom, jointed, the joints swelled, and somewhat flattened, a prominent line runs on each side of the stalk, from one joint to another, and that alternately.
RAMI alterne oppositi, foliosi, cauli subsimiles.	BRANCHES alternately opposite, leafy, somewhat like the stalk.
FOLIA opposita, petiolata, ovata, obtusiuscula, tri-serrata, basi biglandulosa, obtuse serrata, ad lentem ciliata, utrinque glabra, lucidiuscula, venosa.	LEAVES opposite, standing on footstalks, ovate, bluntish, spreading, having two glands at the base, obtusely serrated, if magnified edged with hairs, smooth on each side, somewhat glossy, and veiny.
PETIOLI foliis multo breviores, glabri, supra canaliculati.	LEAF-STALKS much shorter than the leaves, smooth, channelled above.
STIPULÆ quaternæ, ad geniculis, utrinque binæ, minimæ.	STIPULÆ four at each joint, two on each side, very minute.
PEDUNCULI florum masc. axillares, oppositi, erecti, nudi, filiformes, foliis longiores, subtetragoni, superne proferentes glomerulos plures florum, sessiles, odore sambuci.	FLOWER-STALKS of the male flowers axillary, opposite, upright, naked, filiform, longer than the leaves, somewhat four-cornered, producing towards the top, several round, sessile, small clusters of flowers, having the smell of elder.
CALYX: PERIANTHIUM tripartitum, foliolis ovatis, acutis, patentibus, *fig.* 1.	CALYX: a PERIANTHIUM deeply divided into three segments, which are ovate, pointed, and spreading, *fig.* 1.
COROLLA nulla.	COROLLA wanting.
STAMINA: FILAMENTA plerumque novem, alba, capillaria; Antheræ didymæ, flavæ, *fig.* 2.	STAMINA: generally nine FILAMENTS, white and very fine; ANTHERÆ double, and yellow, *fig.* 2.
FLORES FÆMINEI in distincta planta.	FEMALE FLOWERS on a separate plant.
PEDUNCULI axillares, foliis breviores, sæpius biflori, inter flores fœmineos aliquando observatur masculus imperfectus, longius pedunculus.	FLOWER-STALKS axillary, shorter than the leaves, generally sustaining two flowers: among the female flowers we sometimes find an imperfect male flower standing on a longer footstalk.
CALYX ut in mare, nisi quod foliola paulo minora, *fig.* 3.	CALYX as in the male, except that the leaves are a little smaller, *fig.* 3.
COROLLA nulla.	COROLLA wanting.
NECTARIA duo, subulata, utrinque ad latus germinis solitaria, *fig.* 4.	NECTARIES two, tapering, one growing singly on each side of the germen, *fig.* 4.
PISTILLUM: GERMEN subrotundum, didymum, compressum, bifidum; STYLUS sex albi; STIGMATA duo, subulata, patentia, longitudinaliter superne hispida, *fig.* 5.	PISTILLUM: GERMEN roundish, double, flattened, bifid; STYLES scarce any; STIGMATA two, tapering, spreading, on the upper side rough lengthwise, *fig.* 5.
PERICARPIUM: CAPSULA didyma, echinata, bilocularis.	SEED-VESSEL a twin CAPSULE, prickly, having two cavities.
SEMEN unicum in singulo loculamento globosum, extus castaneum, intus album.	SEED one in each cavity, globular, chesnut coloured without, white within.

We can discover no satisfactory reason for calling this species by the name of French Mercury, as it is not peculiar to France, but found with us, in a variety of places: RAY mentions it as growing plentifully on the sea-beach, near Ryde, in the Isle of Wight; and PARKINSON, near a village called Brookelend, in Romney-Marsh, Kent; it would appear to be more common now than formerly, as we very frequently meet with it in waste places, by the sides of roads, and in neglected gardens, in the neighbourhood of London.

The

Mercurialis annua

AGARICUS AURANTIUS. ORANGE MUSHROOM.

AGARICUS *Linnæi G.n. Pl* CRYPTOGAMIA FUNGI.

Fungus horizontalis subtus lamellosus.

Raii Syn. Gen. 1. FUNGI.

AGARICUS *aurantius* pileo conico viscido aurantio, lamellis luteis, stipite nudo. *Lightfoot. Flor. Scot. p. 1025.*

AMANITA glutinosus, flavus, pileo umbonato. *Haller. hist. n. 2420.*

FUNGUS parvus, lubricus, aureus, lamellis raris, amplioribus, pediculo crassiore. *Mich. p. 147.*

FUNGUS aurantii coloris capitulo in conturn abrunto. *Vaillant Bot. Par. p. 67.*

FUNGUS pratensis minor, externe viscidus, striis subtus fulvis seu croceis. *Raii Syn. p. 3. n. 38.*

In pascuis elatioribus solitarius plerumque invenitur, sat copiose nobilicum.	Found plentifully enough with us in elevated pastures, and for the most part singly.
STIPES uncialis, ad triuncialem, nudus, fistulosus, fragilis, et admodum fissilis, crassiusculus, subtiliter striatus, lævis, sæpe tortuosus, plerumque croceus.	STALK from one to three inches high, naked, hollow, brittle, and much disposed to split, thickish, finely striated, smooth, often twisted, and for the most part saffron-coloured.
PILEUS uncialis, aut biuncialis, raro triuncialis, triplurimum sericeus, præsertim in junioribus, lubricus, et subvidesus, primo coccineus, dein croceus, seu aurantius, demum niger; nonulli siccum conicum retinent usque ad dissolutionem, alii plani æout vertice numerorum.	STALK one or two, seldom three inches broad, generally conical, especially when young, slippery, and somewhat chammy, at first of a bright scarlet colour, then saffron or orange-coloured, and finally black; some preserve their conical form even in decay, others become flat with a prominent crown.
LAMELLÆ primo albidæ, dein subcroceæ, si contunduntur statim nigrescunt.	GILLS first whitish, afterwards somewhat saffron-coloured, on being bruised quickly becoming black.

As this Fungus is so distinguishable for its colours, so distinct in its specific characters, and withal so common, it is matter of admiration that we do not find more notice taken of it by Authors. Mr. LIGHTFOOT in his *Flora Scotica* has given an accurate description of it, which cannot fail of making it known: he quotes SCHÆFFER's figure, which represents our plant, and adopts his name of *aurantius*. Mr. HUDSON does not mention it; and we are not certain whether the plant we refer to is RAY be ours or not. As well as Mr. LIGHTFOOT, we had our doubts whether it was the *fragilis* of LINNÆUS; but considering his description, as well as that of VAILLANT, who gives a figure to which LINNÆUS refers, we are certain it must be a different plant. If the *fragilis* of Mr. HUDSON be the *fragilis* of LINNÆUS, it is a very different plant from ours indeed. *Vid.* SCHÆFF. *k. tab.* 230. to which he refers.

This Fungus is by no means uncommon in elevated pastures, particularly where Eye-bright grows. It is usually dwarfish on heaths; but where the grass is not close fed, it is found with a stalk three inches high. The brilliancy of its colour soon strikes the eye. We may observe, that this colour is most vivid, or most inclined to red in the young ones. As it grows old, it becomes yellower, and quickly changes quite black. Indeed it has an extraordinary tendency to turn black, not only from age, but from the slightest bruise. The stalk is also brittle, and very apt to split.

It is found in perfection about the middle of September.

It does not possess any particular acrimony; but is not numbered with such as may be eaten with safety.

Agaricus aurantius.

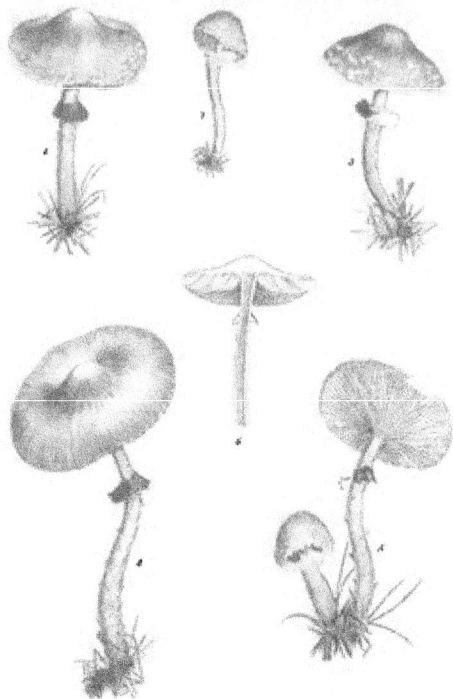

Agaricus aerugineus

Agaricus Æruginosus. Verdigris Mushroom.

AGARICUS *Linnæi Gen. Pl.* Cryptogamia Fungi.

Fungus horizontalis, subtus lamellosus.

Raii Syn. Gen. 1. Fungus.

AGARICUS *æruginosus stipitatus, annulatus, annulo superne nigricante; pileo convexo, cæruleo, viridi, viscoso, lamellis purpureo-fuscis.*

AGARICUS *viridis stipitatus pileo convexo viridi, lamellis albidis, stipite longo virescente.* Hudson Fl. Angl. p. 614.

AMANITA *annulatus, pileo convexo cæruleo viridi, lamellis roseo cæruleis.* Haller. hist. n. 2444.

FUNGUS *medius pileo muco æruginei coloris obducto.* Raii Syn. ed. 3. p. 6. Deering Catal. Stirp. p. 80.

FUNGUS *pileolo cucullato, viscido, intense viridi, et quasi vernigino oblito, inferne lamellis et pediculo albis.* Micheli p. 152.

AGARICUS. Schæf. Icon. tab. 1.

Solitarius, et cæspitosus in sylvis et pascuis colorior, rarius nobiscum.	Grows singly, and in clusters, in woods and pastures, rarer with us.
STIPES bancidus, for triuncialis, ex albo virescens, fistulosus, annulatus, infra annulum floccosus, tores, lubriculus, supra annulum lævis, substriatus, ad infra-fungationem, raro-striatus.	STALK two or three inches high, of a greenish white colour, hollow, ruffled, below the ruffle shaggy, round, somewhat brittle, above the ruffle smooth, and slightly striated, at the half-woolly, seldom perfectly straight.
ANNULUS persistens, tenuis, superne striatus, a fusco nigricans, inferne virescens.	RUFFLE permanent, slender, on the upper side streaked with a blackish purple colour, on the under side greenish.
PILEUS unciam aut duas latus, primo convexo-truncus, ex cæruleo-viridis, lubricus et scabriculus, lævis, prope margine ipsa floccis albidis aspersus, demum planus aut parum concavus, e fusco-lutescens, cuticula facile separanda.	CAP from one to two inches broad, at first somewhat roundish, yet conical, the colour of verdigris, slippery and somewhat viscid, smooth, except near the edge, and on the edge itself, where it is covered with a whitish, shaggy substance, finally flat, or a little concave, of a yellowish brown colour, the cuticle easily peeled off.
LAMELLÆ numerosæ, brevioribus interjectis, e fusco-purpurascentes, parum nebulosæ, demum nigricantes.	GILLS numerous, with shorter ones intervening, of a brownish purple colour, a little clouded, finally blackish.

Amidst that variety of colour observable in the Fungi, there are few in which the green predominates so much as in the present species: hence it affords an obvious character. But, alas! in this plants of a day, we must not lay too much stress on colour; neither so trite a character can be better applied to any subject. It is, however, chiefly in its decline that it loses that verdigris green, which at its first appearance renders it so conspicuous, the cap being often found of a pale yellowish brown colour, and sometimes variegated with green, yellow, and black. The viridity of the cap is so constant a character as its green colour, and that also is most observable in the young ones, especially in the morning, or in showery weather, for in a very dry atmosphere the most vivid Fungi lose their viridity. Next to the greenish viridity of the cap, we may remark, that the edge of it, where it breaks from the annulus, is very apt to be ragged; as have also found, that the inner skin of the cap has an unusual tendency to separate from the flesh. The gills, from the very beginning, are of a purplish brown colour; and the number of ruffle, which connected to the edge of the pileus, receives from the gills a fine powder, which communicates to the upper part of it a dark brown tint; ribs, contrasted with the light colour on the underside, forms a very conspicuous character. The flesh below the ruffle is usually of a blackish green colour, and shaggy.

This Fungus is not very common with us. Several of them appeared this autumn, in a grass plot in my garden; and I have observed twenty or thirty in Earl Mansfield's little wood near the Squared, Hampstead-heath, where, if the season be not remarkably unfavourable, they are with certainty to be found about the middle of September.

It has no smell or disagreeable taste; nevertheless, we do not venture to pronounce it an eatable one.

Ray's description, though a short one, and his referred figure, accord exactly with our plant. Haller quotes Schæffer; we therefore conclude from this circumstance, as well as from the constancy of his description, that our plant is the same as his; and Micheli, who is also quoted by Haller, gives a description so exactly corresponding with Ray's, that we have no doubt but his also is the same as ours. Whether our plant be the viridis of Mr. Hudson, we have our doubts; for he quotes authors who describe two different Fungi; it the same time that he quotes Schæffer, tab. 1. (our plant), and Haller, n. 2444, (our plant), he refers to Micheli, Ray, and Scopoli, who describe another Fungus. Scopoli gives to his the name of viridis; part of his Dispar, is Stirps aeruleus. Ray quotes the Fungus magnus viridis of Sterbeeck, and the viridis an alter aperatus, for in illo receptionf of J. Bauhinus; and Micheli thus describes his, Fungus obsoletus, pileo pallidus, viridi, inferne even pediculo alba. This description is quoted by Scopoli for his viridis. Then it would appear that these two are different species; we must leave it to Mr. Hudson to reconcile their contradictory synonyma.

It could be wished, that every Fungus was as difficult in its characters as the present; we should then have less order lying from that chaos in which the tribe of plants has been involved; as it long involved; not but that those which Linnæus and other Botanists have so much lamented, is rather to be considered as a creature of our own imagination than as the child of nature. The more we look into these variable plants, the more we are convinced that our ignorance of them depends on our inattention and want of observation. Before this issue plants on them as on other plants, observe them in all their states, in all their varieties of situation, and we shall find that each of them has some peculiarity of character. The discovery of this character is what we should aim at; but this will not be found in the closet. We may read over, with the most sedulous attention, Sterbeek, Micheli, Gleditsch, and Haller, or turn over the multitudinous plates of Sterbeek to little purpose; to know the Fungi well we must watch them daily and yearly; in short we must live with them.

AGARICUS CARNOSUS. FLESHY MUSHROOM.

AGARICUS *carnosus* pileo convexo albo, medio rufescente, lamellis confertis albis carne pili duplo angustioribus.

In sylvis acerosis habitat nobiscum rarior, autumno vigens.

Found with us in pine woods in the autumn, scarce.

Solitarius plerumque invenitur, subinde cespitosus.

Is generally found growing singly, sometimes in clusters.

STIPES triuncialis et ultra, magnitudine fere digiti minimi, crassus, nudus, fistulosus, carne diametro tubi, firmus, albidus, saepe rubro maculatus, parum striatus, basi inter folia post emortua descendente.

STALK three inches high and upwards, almost the thickness of the little finger, clumsy, naked, hollow, the flesh the diameter of the tube, firm, whitish, often spotted with red, faintly striated, the base descending amongst the dead pine leaves.

PILEUS uncialis, ad triuncialem, albidus, medio rubescens, et hinc inde maculis concoloribus adspersus, laevis, carnosus, carne multa, solido, albo, primo convexus, dein planiusculus, nec acris, nec lacteolus.

CAP from one to three inches in diameter, reddish in the middle, and here and there blotched with spots of the same colour, smooth, fleshy, the flesh abundant, solid, white, soft convex, finally almost flat, neither acrid nor milky.

LAMELLÆ numerosissimæ, albidæ, angustæ, sesquilineas latæ, brevioribus interjectis, demum rufescentes.

GILLS exceedingly numerous, whitish, narrow, a line and a half broad, shorter ones intervening, finally of a reddish brown colour.

We can find no certain traces of this fungus either in the figures or descriptions of authors; at least in those of our own country. This may perhaps arise, from its being a local, or at least not a common mushroom.

We have hitherto found it only in Lord Mansfield's small pine wood, Hampstead, and there in no great plenty: but having observed them in the same spot, and assuming the same character for several successive years, we are perfectly satisfied of its being a very distinct species. This autumn, Sept. 22, we found about twenty of them.

It is in some degree characterised by the singularity of its colour. We have few fungi that have a white Pileus, with a reddish disk, and that, together with the stalk, irregularly blotched with the same colour; but it is more distinguished by the quantity of flesh both in the Pileus and Stipes. It is this which gives it an unusual degree of firmness to the touch, and has induced us to bestow on it the name of *carnosus*.

Chewed, it discovers no unpleasant taste; but notwithstanding this circumstance, and notwithstanding its tempting appearance, we must, till we have further proofs of its innocence, place it at least among the suspicious fungi.

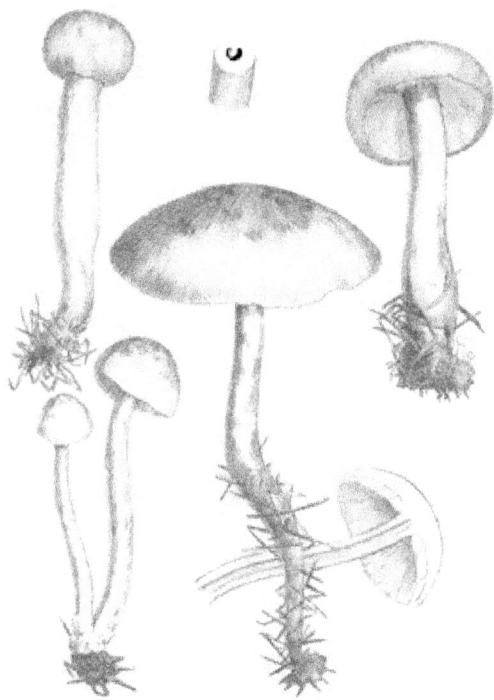

Agaricus vervvvvvs.

Agaricus verrucosus.

AGARICUS *Lin. Gen. Pl.* CRYPTOGAMIA FUNGI.

Fungus horizontalis, subtus lamellosus.

Raii *Syn. Gen. 1.* FUNGI.

AGARICUS *verrucosus* ſtipitatus, ſtipite bulboſo, annulato, mutilo laxo, pendulo, pileo verrucoſo, lamellis albis.

AGARICUS *muſcarius* ſtipitatus, lamellis dimidiatis ſolitariis, ſtipite volvato: apice dilatato, baſi ovato. *Lin. Syſt. Veg. p. 840. Spec. Pl.* 1640.   *Fl. S.* 449.

AGARICUS *verrucoſus* cauleſcens, pileo convexo cinereo, verrucis lamellitique albis.   *Huſſon. Fl. Angl. p. 613. Lightfoot p. 1012.*

AMANITA petiolo poncero fibuloſo annulato, pileo plano ſtriato verrucoſo ſordido lamellis albis. *Haller Hiſt. n. 2397.*

AMANITA petiolo annulato, pileo ſanguineo, lamellis albis.   *Haller Hiſt. n.* 2373.

LEUCOMYCES *gemmatus*.   *Bater. tab. 6. B.*

LEUCOMYCES *ſpecioſus*.   *Batarra tab. 6. A.*

AGARICUS *muſtarius*.   *Scopoli Fl. Carn. n.* 1419.

FUNGORUM *periculoſorum. Gen. 11. Spec. 4.   Claſ. p. 281.   Schaeffer. Icon Fung. t. XX. LXXIV. XC. XCI. CCXLII. CCLVII? CCLXI.*

Solitarie naſcitur in ſylvis frequens.	Frequent in woods growing ſingly.
STIPES palmaris et ultra, craſsitie digiti minimi, ſeu intermedii, ad balin ſemper bulboſus, teres, ex albo-rubeſcens, et maculatus, non raro flaveſcens, annulferas.	STALK a hand's breadth or more in height, the thickneſs of the little or middle finger, always bulbous at its baſe, round, of a reddiſh white colour and ſpotted, not unfrequently yellowiſh, and furniſhed with a ring or ruffle.
ANNULUS magnus, perſiſtens, pendulus, plerumque ſtriatus, ex lamellis impreſsis.	RING or ruffle large, permanent, pendulous, for the moſt part ſtriated.
PILEUS duas, tres, aut etiam quatuor uncias latus, primo ſubrotundus, dein hemiſphericus, demum planus, ad marginem ſaepius obſolete ſtriatus, variè coloris, ſupinè verò non ſordide rubro medio ſaturatiore coloratus, aut diverſicus, plerumque verrucoſus, interdum nudus, verrucis albidis.	CAP two, three, or even four inches broad, at firſt roundiſh, then hemiſpherical, laſtly flat, on the upper ſide, faintly ſtriated at the margin, various in its colour, but moſt commonly either of a dingy red, ſtrongeſt in the middle, or yellowiſh, for the moſt part warty, ſometimes bare, the warts whitiſh.
LAMELLAE numeroſae, brevioribus interjectis, horizontales, primo albae, dein ſordidè carneae.	GILLS numerous, ſhorter ones intervening, horizontal, at firſt white, laſtly of a dirty fleſh colour.

Moſt modern authors conſider the *Agaricus verrucoſus* and *muſcarius* as diſtinct ſpecies. Mr. LIGHTFOOT, ſuggeſts, that they may be only varieties differing in colour. Repeated examination has partially convinced us, firſt his conjecture is well founded; the *verrucoſus* being with us by far the moſt common, we ſhall conſider it as the ſpecies, and the *muſcarius* as the variety: ſo ſingular and ſo beautiful is the variety, however, that we intend giving a ſeparate plate of it.

Before we ſpeak more particularly of theſe fungi, it will be proper to explain to ſome of our readers what is meant by a few terms made uſe of in deſcribing this and theſe or four others, viz. *Volva, Annulus, and Velum,* parts which occur in ſome muſhrooms, but not in others.

There are a few of theſe plants, which, on their firſt emerging from the earth, aſſume the appearance of an egg, and are incloſed in a kind of membranous ſhell or caſe; this caſe we call the *Volva*. If we cut the egg longitudinally down the middle, we may obſerve the incloſed fungus as yet unexpanded. Vid. *Schaeffer Icon. Fung. tab. 244. fig. 1, 2, 3.* As the muſhroom proceeds in ſize, it burſts open this *Volva*, and ſometimes leaves it broadly behind, as in the *Phallus impudicus*; but more frequently the upper half of it is borne upwards on the Pileus, or Cap, which, not being ſufficiently large to cover when the *Velum* is expanded, it breaks in various directions, and appears in the form of a number of little knobs or warts irregularly ſcattered. Such then is the origin of the warts on the membrane, which ſeems them may ſometimes be diſſolved dint ordinary; or as it may be rubbed off in the muſhroom pullis, either out of the ground; or deſtroyed by heavy rains, or other accidents, ſo we never find theſe warts alike either in number or ſhape in any two fungi, and frequently entirely wanting; but if no extraordinary accident happens, they will be found in every artificial-raiſed fungus of this ſpecies. We may remark, that the *Volva*, which we have thus deſcribed, is not the *Volva* of Linnaeus his *Velum* is our *Annulus*.

In many of the fungi the gills are covered and protected in their infancy by a membrane, more or less thick, totally independent of the Volva, attached to the edge of the Pileus one way, and round the stalk the other. While the membrane is visibly thus connected, which is just as the Pileus is beginning to expand, we call it the Velum or Veil, though generally the term is applied to those membranes which are remarkably thin, almost like a cobweb, and which, when the Pileus is expanded, leave little or no traces of their existence behind, as in the *Agaricus fasciolaris*. The greatest part of this membrane in *separating* is generally left either with the Pileus or Stipes: sometimes what it leaves remains with the Pileus, and is only sufficient to give the edge a ragged or toothed appearance; but more commonly, where it is in any degree substantial, it leaves the Pileus, and attaches itself to the Stipes, where it either progress horizontally, as in the *arvensis*; or becomes **pendulous**, as in the present species. This part, thus attached to the stalk, we call the *Annulus, Ring or Ruffle*.

There are three characters which distinguish the perfect species of mushroom, viz. a cap, more or less covered with warts; a stalk, bulbous at its base, and furnished above with a pendulous striated ruffle. These will be found in every perfect fungus of this sort. Colour is not to be depended on; the cap being sometimes, as in the variety *muscarius*, of the most beautiful crimson, and very frequently, especially in Charlton Wood, of a cream colour; but its most usual tint is a dingy red, inclining to brown. The Gills are always white at first, and become of a dingy red at last. The stalk in those which have a reddish Pileus is usually mottled with red and white. The whole fungus, but particularly the base, is apt to be soon destroyed by the larvae of various insects, and among others by those of an undescribed species of *Tipula*, somewhat less than the *Tipula plumosa*, and distinguished by having its legs unusually hairy. It was by accident we discovered the attachment of this insect. Between the Velum and the Gills, previous to the separation of the former from the edge of the Pileus, there is a considerable cavity. In this cavity we found, in a young fungus of this species, at least twenty of these Tipula, which had introduced themselves through an accidental aperture in the Velum.

The *Agaricus verrucosus* is very common in all our woods about the middle of September. The *muscarius* is plentiful only in particular spots.

We had the curiosity to taste this showy fungus. Chewed, it was not unpleasant in the mouth; swallowed, it quickly produced a disagreeable burning kind of sensation in the throat, which extended to the stomach, though the quantity swallowed was but small; and this sensation continued a considerable time. That I might not be mistaken in my idea of this sensation, I prevailed on my draughtsman and gardener to chew and swallow some of it, who complained of its producing a similar effect. Hence we may infer, that this species, taken in any quantity, is likely to prove highly poisonous. This effect accords with the account given of it by different authors. Scopoli makes mention of some persons being poisoned by it, mistaking it for the *Agaricus caesareus*. Haller relates, that six persons of Lithuania perished at one time by eating it, and that in Kamtschatka it had driven others raving mad; that there, three or four of them are eaten without much effect, but that ten intoxicate: nevertheless, the Russians eat it with their food; and the inhabitants of Kamtschatka prepare a liquor from this fungus, and a species of Epilobium, which, taken in small quantities, inebriates, and produces a trembling of the nerves, making some joyous, others melancholy. The very urine of those who drink it is found to intoxicate. Linnaeus says, that flies are killed, Scopoli only stupified, by taking an infusion of the *muscarius* in milk, whence its name, and that it is also inimical to bugs; but we have certainly much better remedies for these troublesome insects.